THE
WORLD
& US

★ THE ★
WORLD & US

★

Don Clark

CONTENTS

INTRODUCTION

This book is about the world and America's involvement in it. Primarily it is a selection of newspaper columns I wrote between 1981 and 1985 focusing on the major issues of international affairs: the arms race, world hunger, Central America, the Middle East, terrorism, the Soviet Union and the USA, and ethical questions of military involvement and national policy.

The columns were not written to make money or to persuade readers of the rightness of the writer's views. They were written to interest Americans in the subject of world affairs, to raise their awareness of the complexity of the issues, and to get them to ask questions, seek answers for themselves, and let the policy makers know what they expect of them.

The columns were a direct result of a US Presidential Commission Report sanctioned by the Carter administration. The report, entitled STRENGTH THROUGH WISDOM, suggested that the general knowledge held by Americans about the rest of the world was, in a nutshell, abysmal. Further, its members concluded that the USA was doing itself great harm caused by this ignorance of the cultures and goals of other nations and the tendency to treat them as if they were, or wanted to be, carbon copies of the USA. So much so, they argued, that the creation of programs to inform Americans about the world would be more valuable to the success of our foreign policies than developing new missiles or other weapons of war.

The Commission proffered many recommendations to help deal with this ignorance, almost all of which were promptly ignored, but among them was the suggestion that if we could get more Americans to read regularly about world affairs it would do much, if not more than anything else, to overcome our apparent and tragic misperceptions of others.

That comment struck home. I had moved to Bozeman, Montana, and the university atmosphere there from Washington, D.C., and a stint in the Office of International Negotiations of the Joint Staff, Pentagon. The change had been enormous and much of it represented by the difference of world news coverage readily available to me in DC versus the Montana papers. Coincidentally with my reading of the commission report I had kept a business luncheon date with a local private air service president. Our conversation had turned from business to trivia and then to his questions about my life in the USSR as a military attache, my experiences negotiating with the Russians, and my classes about that enigmatic Red nation. My comments surprised him, as they had thousands of people in audiences all over this land during the then almost ten years since I had served at the US Embassy. Indeed I had long ago realized that my current thoughts about the USSR and US/Soviet policies would have surprised even me before my extended stay in the USSR and the subsequent years of research and study of this pretender superpower. Knowledge about that subject, as most, had wrought a change in perception.

"Why haven't I heard these things before?" he queried. When that question was stated and merged in my mind with the just read commission report, an old desire of mine to be a writer was resurrected. (I first felt the thrill of being considered a writer in about the third grade when my teacher put a short story of mine, entitled "The Prone Ranger," on the bulletin board for a parent's night.) Thus, a new columnist was born.

I initially called the column "Hither and Yon" to illustrate that the emphasis would be on the rest of the world, yonland, but primarily in its relation to US (the USA, you and me, the Hither). I focused through the years on the USSR because I know it best of all the rest of the world and because it is America's greatest rival as the most influential nation on this globe. And I focused on US/Soviet issues like the arms race for obvious reasons. But I also tried to write about far away and seemingly less important places like Timor, Eritheia, Baluchistan, Turkey, and Panama. It was a weekly treat for me to decide if some world issue was so pressing it demanded the space or whether it was an appropriate time to introduce my readers to a problem or place they were not likely to read about without turning to the low circulation professional journals. Hither and Yon as a name never caught on with my editors so for this book I have chosen the title of a companion weekly radio show I called "The World and US." In the title the "US" part has a double meaning standing both for "us," you and me, we the people, and for the USA," our collective nation and government.

It worked. The column appeared at one time or another in four Montana dailies and developed quite a readership. Several of the editors told me they received more comment, both pro and con, on my column than any they had ever carried, local or nationally syndicated. The column was also selected by Voice of America as one they would air excerpts from on their overseas brcadcasts. That was pleasing as was the receiving of the cards, letters, and phone calls. But the most heartwarming result was to watch each of the papers involved slowly increase the space they devoted to international news. I am sure much of that was the natural evolution of their papers and the times, but I have no doubt that the response to my column helped illustrate to the editors and publishers that Montanans did care and want to know about what was going on out there in the world beyond its paradise-like piece of the globe.

The column led to many interesting experiences, trips around the USA and Canada as a speaker on arms control and US/Soviet issues, a rare return of a former US military attache to the USSR without the protection of a diplomatic passport, the doubling and tripling of the sizes of my classes (a mixed blessing), phone calls in the middle of the night (some pleasant and some irate), and interesting mail. My favorite letters came from a mother in Butte who was shocked to think that a "Red lover" like me was teaching in the Montana college her son attended. The next letter was worse as she discovered he had taken a class from me and loved it. The third represented quite a change. She had had a long talk with her son, decided I was an all right guy after all, fair in offering many views to my students and in getting them to think for themselves. She was now a fan and even wanted to know if she would be welcome to take one of my classes. I invited her and regret it never worked out. But clearly not all who disagreed with me later became supporters.

There is one fan, I use the word loosely, who continually looks to my education by sending me clippings, primarily from the *Reader's Digest*, which he assures me prove that the USSR is the most vile, untrustworthy, heinous of nations this world has ever known, and that I have been duped when I suggest we could ever reach some kind of a modus vivendi with them. Better with the devil he suggests. I tell him I am glad he and the president have the same sources of information, but that I prefer far more detailed and authoritative sources.

Probably the most emotional encounter resulting from the columns occurred when a couple of very conservative local groups brought in their "hit man" to challenge me to a debate. He was a Soviet defector, in fact a self-admitted former propagandist for the communist party. Since few in the audience that night other than myself had ever listened to a Communist party "agitprop," I suspect I was the only one there who realized he was giving just about the same speech to them he must have given to thousands in the USSR in his earlier days. In his new version he just changed the name, "USA" to "USSR" when he got to the villainous parts of the speech.

But it was quite a night. A crowd of about 800 flocked into a room set up for 300 and his sponsors came early, packed the front middle and were there to cheer the champion of their viewpoint. I knew it was going to be a difficult evening when in the first two minutes of his initial presentation while describing his checkered past under communism, he rose to his tiptoes and shouted out boldly and proudly words to the effect, "But now I am a red, white and blue American and proud of it." The crowd response was a great cheer, as if on cue, and one or two even jumped up and began doing a kind of war dance in the aisles.

Allegedly, we were to debate what would be the best US foreign policy to pursue in dealing with the USSR, but he seldom touched on that preferring to spew out words of hatred and disgust with liberals, congress, the CIA, former presidents and the like who might have favored detente, arms control, or the cancellation of a US weapons project. All such folk, he declared, were traitors or dupes playing into the hands of the villainous Soviets, who of course were not real Russians, as he apparently was. I found it ironic that he was able to leave the USSR and win this chance to earn a living as an entrepreneur as a result of detente, but his response to that was that he would rather have stayed in the USSR than have had the USA hoodwinked into thinking the Soviet leadership might be human.

But as unpleasant as the evening was, it was a worthwhile one also. For many in the audience reported to me and others that they learned a great deal about intolerance, about emotion and hatred versus reason and discussion. Many students sought me out over the next several weeks to tell me that for the first time they understood how Hitler had swept the Germans along into Nazism and the holocaust via hatred wrapped in emotion and blatant appeals to human superiority preferences. Objective viewers assured me I won the debate on points of logic, arguments, data, etc. On the other hand, the noise level clearly favored the defector. But to those who went there with an open mind, I think they saw and heard a good example of the contrasting positions, the traditional view which says the Reds are devils and can only be contained by superior power, and the non-traditionalist view that argues superiority is an obsolete term in our current nuclear world and that the US/USSR can, and indeed must, find a modus vivendi to control the arms race and coexist.

The chapters that follow often fit the shape of that debate. For basically the essays that follow argue that the USA leadership, primarily because it does not well understand the real world it exists in, be it the USSR, Central America, The Mid-East, Europe, or elsewhere, is engaged in foreign policies which are neither very successful nor worthy of a nation as great as ours should or could be.

In the brief introduction to each chapter, I will relate what triggered me to write them and how the public seemed to respond. Where new events have occurred to merit additional commentary I will add a previously unpublished column to bring the issue up to date. I ask each reader to read each column and chapter critically comparing what is said here to the conventional wisdom prevailing in the USA on the subject. Do not read these to agree, or even to disagree, but rather to pursue a thoughtful examination of the issue. When finished if doubts have been raised, "voila!" But if not and your previous convictions have merely been reinforced by the rethinking process, "wunderbar!" For there are no right and wrong answers in international affairs, merely opinions, hopes, informed guesses.

In the process of reading the following chapters if your appetite is whetted to know more about the terribly important issues discussed herein then consult the bibliography at the end and read to your heart's content. For that has been the motive of the whole exercise, to get readers to think more about the world, its problems and the differing peoples in it. ENJOY THE ADVENTURE!

CHAPTER ONE

The USSR

When I originally planned this book the first chapter was to be about the arms race. But as I was doing the editing, I remembered how often in my class discussions on that topic students came to the conclusion that the arms race simply is not resolvable until the leadership and citizens of the USA and the USSR develop more realistic understandings of one another. It is a conclusion reached by many who have delved into this international field. Roger Mollander, a former member of the Nixon national security council staff and the founder of Ground Zero, reached that same conclusion a few years ago and wrote to tell me that he had founded The Roosevelt Foundation dedicated to that very task, educating Americans about the USSR so that eventually the arms race madness might be manageable.

Thus, I pulled a switcheroo and placed this chapter on the USSR first to be followed by that very important subject of the arms race.

In the years since I served in the US Embassy as an Assistant Air Attache, I have found enormous interest among Americans about their counterparts in the Soviet Union. But just as the interest is high so is the level of misperception. And strangely that is also something I found among Soviet citizens when I have been there, a burning curiosity about the USA and a concomitant mythology about us that to an American seems positively absurd. It is little wonder that our nations mistrust, fear, and miscalculate the actions of one another, for the abyss of ignorance that separates us is huge. If great powers like the USA and the USSR are to avoid an Armageddon then understanding is vital. Understanding does not mean agreeing, but simply a better perception of what your actions might mean to them and vice versa.

Critics of my column have often called me a Soviet apologist, a Red lover, or a dupe of soviet propaganda. I deny the first two wholeheartedly and doubt the third. Any careful reader of my columns would have to admit that I am often critical of the USSR seldom passing up a chance to point out that Soviet communism has proven itself to be the third best of the three viable economic models the world is practicing. I often note that the Soviet system is seriously flawed by an overconcentration of power into the hands of a very few who are inevitably corrupted by that power, and I denounce their control of East Europe, their occupation of Afghanistan, and the checkered years of Stalin, a man whom I often call the champion murderer of all time.

But readers get that apologist idea because I also try to write objectively rather than emotionally about the USSR. I try to point out their economic progress, their modernization, the enhanced living standard in the non-Slavic republics which are most remarkable when compared to their ethnic brothers outside the Soviet borders in places like Afghanistan, Iran and Turkey.

I also try to help readers understand how the Soviets look at the world, to explain the rationale for their "paranoia" about defense and secrecy, their preference for strong leaders, their disdain for some democratic principles that we, somewhat uniquely on this globe, consider as international values although they are not. To show how Soviet people do have some say over their lives, to note their enhanced "wiggle room" in employment, housing, influence on their leaders, etc.

These are apparently not popular subjects for they stray from the more typical American media barrage about the USSR that paints it in dark colors and of course makes us look so much greater by comparison. But I feel there is enough of that so I go out of my way to find ways to proffer examples about the USSR that point out why it survives almost 70 years after the Red takeover without daily rebellion. Why it has supporters, workers and even admirers, although admittedly there are fewer and fewer of such in recent years as the Red economic progress slows and compensates less for its human rights mistreatment.

But even by human rights scales the USSR of 1985 is not the nation of 1935, '55, or even '65. Once millions disappeared in the middle of the night for often imaginary crimes, never to be seen again. Today thousands may still be political prisoners, far more than can be justified by any measure, but they are handled quite differently. Arrests usually come after ample warning and repeated efforts to persuade the dissident to be quiet and follow the norm. Now there is usually a trial, albeit not a very fair one, a set term, a known prison location, and visitation and letter rights. Hardly a pretty picture but far more manageable than the reign of terror under Stalin and his purges. OK? Of course not, but changing and better, yes!

The USSR I describe in the following pages is different than what most of you readers suspect. I doubt if any of you would prefer to move there, I certainly would not. But it is a USSR we need to know if we ever expect to carve out a modus vivendi with them. And for the sake of the world, like it or not, we need to do just that. If the USSR were actually like that place Ronald Reagan describes, we could and should not, but if it is really like the one I think I have seen and men like George Kennan describe, we can. You decide!

1/30/83

USSR Inc.

Premier, Chairman, and President. The USSR, unlike most nations, has them all. And a lot of people get very confused by that and other news out of the USSR since that nation follows a different system of economics and government. Perhaps the following will help you keep things straight.

Think of the Soviet Union as a giant corporation. Such corporations have governing boards and a Chairman of the Board whom everyone agrees is "numero uno." The board, under the guidance of the chairman, makes all of the major policy decisions for the corporation. In this USSR analogy the board is the Politburo of the Communist Party, and as Lenin said way back in 1920 or so—" no important decision is made in the USSR without the approval of the Politburo." The newly elected Chairman of the Politburo is Mr. Andropov. The chairmanship makes him the commanding "commie." But every corporation must have managers as well who deal with those very important day to day questions of operations of the organization. These people make decisions and monitor progress, but they do so within the policy guidelines of the Board, and seek the board's blessings when things arise that are not covered by policy. In the USSR the equivalent of that managerial staff is the Council of Ministers. The number one man on that council, a sort of Chief of Operations, carries the title of Premier.

For years that position was held by Alexi Kosygin and we in the West often wished that this seemingly moderate technocrat could secure the role of number one. We speculated that he might since he was virtually in charge of all the production of goods and services in the USSR. In corporations sometimes Chiefs of Operations do get promoted to the Head of the Board, but seldom does that happen in nations. Premier Kosygin and his current replacement, Tikhonov, might play a key role in the USSR, even sit on the board, but they are technocrats first and politicians second, and in nations politicians almost always have the final say—they certainly do in the USSR. Tikhonov's operation attempts, often futilely, to control all of the goods and services of the USSR. There are over forty thousand such enterprises from soldiers to tractors and computers to bobby pins. Most are not really under control and often flounder in a maze of black market and bribery.

The USSR has another organization called the Supreme Soviet. Most of Americans equate it with our Congress since, according to the Soviet constitution, it is the highest legislative body in the land. But that comparison can be terribly misleading. In our corporation example, I think the Supreme Soviet would be best represented by the role of corporation salesman. Corporations need salesmen and generally reward their best with titles, recognition and often even trips to Bermuda or other exotic locations. The Communist leadership of the USSR does much the same. By party control of the nomination process for election to public office, they reward those Soviet citizens who best have helped sell communism by arranging for them to be elected to the Supreme Soviet and thus receive short trips to Moscow. This, like a "Salesman of the Year" award, proffers nice recognition but does not necessarily grant its recipient additional power. Although the SS meets for 2-4 weeks almost every year and has for many a year, they have never ever voted "Nay" on anything that has ever been proposed to them. It is certainly not like the US Congress!

Since the SS is in session so seldom it elects from its membership a committee that acts as the legislative force when the whole body is not in session. The group is called the Presidium and the head of the Presidium is given the title of President of the USSR. Currently that office is vacant. Brezhnev held the title along with his party chairmanship as did Khrushchev for part of his tenure and Stalin for much longer. Podgorny, who just recently died, held the office for much of the Brezhnev reign until Brezhnev squeezed him out and established his joint tenancy. Most believe that Andropov will not be indisputably the boss of the USSR until he too acquires the trappings of that second title of president. But it is really no big loss to him if he does not do so immediately. When Nixon and Brezhnev signed the SALT 1 Accords Brezhnev was not yet the President of the USSR, but no one suggested that Podgorny should be the one to sign the agreements and drink the toasts with the American president at the auspicious moment. Russians know who really runs their system.

Many of the top leaders have arranged to serve on both the Politburo and the Presidium. This corporation style of leadership and direction from above is why many call the USSR's system, "State Capitalism."

7/18/82

The Soviet Jigsaw

These days many of our leaders are telling us confusing stories about the USSR. On the one hand Kissinger, Haig, Richard Pipes of the NSC, even Reagan himself are all quoted pointing out Soviet economic and political failures (a declining GNP and productivity growth, failed consumer industries, and Poland's revolt), while recanting on the other, the enormous threat of Soviet power (the Afghan invasion, the SS-20 missile, the window of vulnerability, etc.). It's all very confusing; are they Red monsters who threaten the world, or are they a decaying empire headed for the ash heap of history?

As usual the answer lies somewhere in between the extremes. We have a tendency to overrate both the USSR's successes and failures. What the so-called experts mean to say is that the vaunted economic successes of the first communist nation were garnered at a terrific cost in human bloodshed and sacrifice and in a society whose economic takeoff point was very very low, but now the well has run dry. The Soviet economic model turns out to be, over time, only mediocre but viable.

They have a huge military/industrial complex that makes them a developed nation, but in the consumer and agricultural areas they qualify only as a developing state. There is no unemployment but considerable under-employment as they create make-do tasks in order to find jobs for all. The central planning idea has been overwhelmed by the complexity of modern society and the amount of waste, confusion and inefficiency is appalling. Recently this has been borne out by both a detailed UK study by thirty specialists on the USSR and confirmed by a *Pravda* article by none other than Vadim Trapeznikov, a former Deputy Chairman of the Soviet State Committee for Science and Technology. Both refer to more problems than successes in the Soviet economic model.

While the West seems to cycle between bull and bear in its economic results, the Reds fluctuate between mediocrity and scarcity. The central planning idea was thought to be able to eliminate such ups and downs and to take care of people's need instead of greed, but it has not worked out as planned.

All of this then raises the question, how can the Soviets be such a military threat to the West? My answer is that they are not! They are not only not ten feet tall, as often portrayed, but better classified as about five foot seven. Their military force is large and formidable, but like the economy its weaknesses for modern warfare far outnumber its strengths. They are capable soldiers; I would certainly not suggest that the US and our allies deliberately attack or provoke them. In Europe they have a conventional edge, although probably an insufficient one to take advantage of, while overall we have an equally unexploitable slight nuclear superiority. This status is well described in a superb new book by Britain's Lord Solly Zuckerman entitled *Nuclear Illusion and Reality*. Theodore Draper calls it the best book yet on the nuclear issue, and I agree that it is worthwhile reading.

So in my opinion to excessively fear the Soviets, as many of us do, is foolish. Their equipment is generally inferior (as revealed frequently in Arab/Israeli battles), their tactics less than revolutionary (as illustrated by their frustration in Afghanistan), and their leadership best described as "political pygmies" (the description recently used by the highest ranking Soviet defector to the West).

But saying all of this can also be misleading. It does not mean the USSR is on the verge of collapse or revolution, or about to launch a desperate war in order to stave off one, the other, or both.

In reality the Soviet Union is a medium tier nation state, a nation perhaps rivaling Italy or about to pass Great Britain in its success at providing for the economic needs of its people. But it has brought those people from a position almost a century behind the living standards of the best off in the world to only about twenty years behind. Simultaneously it has become *a* but not *the* military power of the globe.

Recognizing this state of affairs a better informed US should be able to deal with the USSR more effectively. I suggest we should move our competition as readily as we can out from the current military focus to the broader territory of ideals and economic competition where we hold the best cards. We can risk doing that only if we conclude, as I believe the facts truly support, that their military threat is considerably overrated and especially so in the area of nuclear weapons, the only ones that could truly endanger us.

I wish that every American could visit the USSR for a few weeks and that every dissatisfied non-US soul on this planet could have a free vacation in both the Soviet Union and the USA. If that were to occur, the over blown image of a great and powerful USSR would dissipate and the inclinations both to copy its system or to endanger ours by exaggerating their power would significantly decline.

But why wait for that unlikelihood? The facts are available for all to see and only sometimes hidden by the misleading statements made for reasons of polemics. The USSR is there; it is not about to fall, indeed it is even getting better and stronger, but it is still not, and is not likely to become, a valid threat to the USA unless we were to unilaterally disarm. So let us get on with the perfecting of the American dream and cease wasting excessive amounts of our precious resources on futile efforts to ward off the bogey-man, who like all bogey-men does not really exist, except in our imagination.

———————————————

Vive La Difference

Aren't all people really alike? NO, my training and experience have taught me that the diverse cultures of this world have produced people who think and act quite differently, even under similar circumstances. This "la difference" in people causes a great many problems which can all be lumped under the heading, "lack of understanding."

Let me illustrate using the Russians as the example. If the following seems to be critical or to brand Russians as of a lesser quality—it seems so only because we will be judging them from our cultural perspective rather than from theirs.

Russians prefer leaders who are not just strong but demonstrably powerful. Their attitude seems to be that no one wants to be led by ordinary folk; a leader needs to prove his mantle with boldness. Knowing this helps us to understand why men like Stalin, Khrushchev, and Brezhnev, with their records of iron fist rule, are tolerated by Russians when people like us would rebel. When living in the USSR I asked tens of Russians whether they preferred Stalin or Khrushchev as a leader. All Russians are aware that Stalin killed millions of his own people while Khrushchev exposed Stalin's crimes, loosened the screws perceptably, and even acted at times like a western politician, playing up to the people as if seeking votes. Yet, almost 100 percent of those asked, replied, "Stalin." To my incredulous, "why?", they usually responded that Khrushchev was merely a country bumpkin whom no one could admire, while Stalin was a man whose mere glance could make one shudder—a powerful, awesome, and clearly cruel man. They nearly universally hated him, yet; held him in great respect due to his immense power and boldness in using it.

This preference for omnipotent control shows up in another way that seems strange to most of us. As a group Russians are not bothered by interference in daily life by government, indeed Russians seem to expect instructions from on high and feel that they make life easier. This phenomenon was vividly demonstrated to me one morning while a Canadian and I were walking along a Soviet beach along the Black Sea. Groups of Soviets passed us going in the opposite direction, and they stared and pointed at us as if we were doing something wrong. After checking to ensure that our zippers were zipped and all in proper order, we became puzzled by all of the attention we were attracting. Finally a group of Soviets stopped and one old lady said, "What's the matter with you, why are you walking in the wrong direction?" She explained to us that the government had placed colored arrows along the beach to point out the best direction to walk for morning or evening views. We were erroneously walking in the evening direction instead of the morning one, and we stood out like what we were, strangers in a directive society.

Andre Amalrik, the Russian dissident writer who died in a car wreck on his way to the Madrid review of the Helsinki Accords, (a meeting at which the West roundly criticized the Soviet human rights record), understood this difference of Russians and pointed out numerous examples of it in his book, *Will the Soviet Union Survive Until 1984?* One classic example he illustrates is how Russians react differently from Europeans and Americans when a neighbor or a friend seems to be succeeding beyond expectations. Although many in both cultures may have the same human reaction of jealousy, Amalrik suggests that the Westerners then begin to consider what they can do to match the success of the friend. Does it require better contacts, more education, a willingness to lie, cheat or steal, work harder, marry the boss's daughter or what? The goal becomes that of somehow achieving equal success. But for Russians, Amalrik posits, the reaction usually will be different. They begin thinking what is it that they can do to bring the more successful one back down to everyone else's level. Amalrik's explanation is that those of European and American backgrounds have evolved farther beyond peasantry than Russians. He says one can merely scratch the surface of any Russian and find a peasant underneath. Peasants have a zero-sum world concept in which one person's gain can only be achieved at everyone else's loss. Thus the gains of all others are to be resisted. There is only so much of all the needed ingredients of life, and when anyone begins to acquire far more than is needed, he is depriving the rest unfairly and must be brought down for the good of the majority.

Such an explanation helps one understand the Russian emphasis on secrecy and security, a fetish that existed long before communism came to that land, and which may have helped make it such a ripe place for such a new order. This cultural predisposition helps explain why Stalin and his successors have all given such a high priority to building the strength of the USSR first, even at an enormous cost in suffering, and why the people have accepted that approach.

I believe it true that all of humankind share certain common traits such as love for our offspring, passion, etc., but we really have far more differences than are generally recognized and those differences can cause a great many problems if not recognized. "Vive la difference," but only if we have the wisdom to recognize that it exists.

11/25/81   # Stalin's Russia is Gone

Almost every objective observer is now willing to agree that the USSR was a most unpleasant place to live during the reign of Stalin with his tens of millions of executions or deaths by sentence to hard labor in the cold storage of Siberia. But many people ask me, "What is the USSR like today? Is it as terrifying as under Stalin, as drab and without joy as we in the West are led to believe?"

I am happy to report that, Solzhenitsyn's accounts aside, the USSR is a better place to live during the '70s and '80s than in the '50s and before. Mind you, I am not suggesting it is great or even would be considered acceptable by those of us who have lived in the West and under free enterprise; it's just better now than it was under Stalin or the czars.

Let me illustrate my point through the mouths of Russians. A young Russian Intourist guide, whom I became acquainted with via my many visits to the airport awaiting special American visitors, "roughly" put it this way. "Oh it is much better now," he said. "We used to lie awake fearfully at night and await the sound of footsteps in the hallway or the knock on the door in the middle of the night. If you heard such sounds first you feared for yourself and family and then for the neighbors. If it was not your door they stopped at, you wondered all night which neighbor would be missing tomorrow. Those that were taken usually just disappeared. No one knew where or why and other than rumor, they might just never be heard from again. It's different and better today—THEY KNOCK ON THE DOOR IN THE DAYTIME."

When he spoke that last sentence, I was startled and thought that he was either putting me on or telling me that really there was not any difference. But after further discussion it sank in, and I realized he meant precisely what he said. It is different and better when they knock on the door in the daylight. Any trauma is easier to face when it is done in the clear light and without the added fear of the unknown. Although in much smaller numbers, people in the USSR are still arrested and sent off to camps, but today everyone knows who is likely to be arrested and where they have been sent. Relatives and friends may receive mail from the prisoners and can even visit them. They are often released, sometimes in exile, within five to seven years.

Russians would also argue that today those who are arrested bring it on themselves. The rules today, unlike in the past, are clear, and it is not a careless word or an unwitting mistake that leads to one's arrest. It most often comes to those who are so fed up with the system and its shortcomings that they openly defy it and almost invite arrest after ample warning and cajoling by the party and the secret police.

Material life has improved even more. The communist system, with all of its shortcomings, failures and inefficiencies, can produce a bare sufficiency of those products that modern populations need to be minimally happy, especially in a country that has as much natural resources and labor available as the Soviet Union. Most Russians can now afford acceptable clothing, food, drink, heat, a TV, radio, and a place to live. They pay for most of those items with many more hours of labor than we in the West do, but unlike under Stalin, today, few in the USSR freeze, die, riot over hunger, or complain about a lack of adequate living standards.

A girl named Tonya made this clear to me. She had learned of Svetlana Stalin's defection (Stalin's daughter, who years later returned to the USSR) even before the Soviet authorities had revealed it, and was fascinated by the story. I loaned her Svetlana's first book which described the defection decision, and on several occasions I translated for her Svetlana's comments as reported in the Western press. After one such translation in which Svetlana, to paraphrase , indicated she defected in order to live her own life, to go where she wanted to go, to say what she wanted to say, to write and read what she wanted, and to be able to worship freely, Tonya shook her head and with a perplexed look on her face said, "Why would anyone want to leave Moscow, we have electricity and heat." That was 1968 and Tonya, although quite young, could well remember when even Muscovites, who live in the richest city in the USSR, were wanting for electricity to read and cook by and heat to warm by on those cold wintry nights.

"I'd rather be dead than Red." That slogan is popular in the West, but it simply does not wash with most Soviets. Such statements are reserved for those of us who have long had the necessities of life in considerable abundance, and thus the luxury to complain about the more subjective limits on life. Those thoughts are often shared by artists as well, writers, poets, musicians and the like whose great creative skills are squeezed by the tight Soviet society. But most people, and indeed most of those who live inside the Soviet Union's borders, do not really feel that way. To most of them communism has provided for their materialistic needs better than any previous regime while simultaneously leading Russia into a semi-superpower status. Something that flatters their ego and makes them proud to be the first great communist nation. This majority would counter, "Red is better than dead, destitute, or divested of job, health services, education, etc." Their life may be bland, but it is not totally empty, especially when compared to their recent past.

10/19/80   # Why Afghanistan?

During the last several months the question most frequently asked me has been, "Why did the Soviets invade Afghanistan?" The query is usually prefaced or followed with a remark such as, "Is it because America has grown so weak, or are the Russians planning to seize control of Iran or our middle eastern oil?" To answer this question we need to look at the issue through Soviet, not US, eyes. The Russians have generally been quite conservative about the use of force outside of the Soviet Union except under one very special circumstance; when they see a threat to the stability of the USSR or the commonwealth (USSR plus East Europe) itself. Does Afghanistan fall into that internal threat category? It does, because of a new and not so widely known development. The most recent Soviet census revealed that Russians are about to become a minority in their own union. All of the many non-Russian peoples within the Soviet Union's borders now number just about half of the population and they are growing at a faster rate than their Russian overlords. Soviet Moslems are the largest of these non-Slavic minorities; they compose about 20 percent of the total population, are the fastest growing group and are the majority population in regions bordering on Iran and Afghanistan.

This minority problem has been an issue of great concern for all of the almost seventy years of communist rule. I suggest that their concern for this minority problem was their primary, but not sole, reason for the Afghan invasion, and that certain coincidental circumstances came together to make the military option probable in 1979.

For several years prior to December 1979, Afghanistan had experienced a series of military coups, supported and encouraged by the USSR, which had made the Afghan government socialist proclaimed, close, cozy, and dependent on the Russians—a goal not at all new to the communist czars, but in fact one that had been sought by *all* previous governments with varying vigor and success. But the news was not that the Afghan governors were finally chummy with the USSR, the news was that those friendly leaders were in deep internal trouble. The government was clearly losing a battle to a band of rebels. Again, it was not unusual in Afghanistan for rebels to be in conflict with the central government in Kabul, but what was different in '79 and today is that the rebel groups were winning converts rapidly. Their rallying cry was and is their religion; they labeled their war a "jihad," a holy war, and called themselves "mujahidin," Islamic holy warriors.

Couple this unusual unity and success of those rebels against their central government with the fall of the Shah of Iran, Khomeni's incredible succession, and his cries for a united Islamic Empire opposed to all heathens (and no one can be more heathen than the self-professed atheists/communists of the USSR), and the success of a Moslem movement on the USSR's borders would have to appear to the Russian leadership as a genuine threat to their control of the Muslim territories within the USSR. I suggest the Soviet leadership saw it that way and concluded that they could not afford to let those Muslims inside the Soviet Union observe an Islamic oriented revolution succeed in overthrowing a Soviet supported socialist government. Probably no national decision is made on a single circumstance alone. But the hard evidence indicates that the main reason for the Soviet invasion was the same one that led to their earlier attacks on Hungary and Czechoslovakia—an abnormal fear that a successful small resistance would spread like a plague and infect the enslaved millions within the commonwealth and the USSR itself. Since the communists took over in 1917, when *they* represented only a small fraction of the peoples or even the politicians of Russia, they have had a paranoia about the ability of a small number of dissenters to overthrow them.

But we must return to the additional clauses attached to the opening question; was it America's weakness or do the Soviets plan to take Iran or seize control of the middle eastern oil? I usually respond to these queries with questions of my own. What are the points cited by America's defense critics to exhibit America's claimed new weaknesses? The leading answers are our failure to develop the B-1 Bomber, the MX missile, the neutron bomb, or a rapid deployment force. Now ask yourself, reader, would the actual operational deployment of sufficient numbers of any one or all of those items, regardless of the pros and cons concerning their real value to US forces, have stopped the USSR from invading that backward, barren land located in their backyard, thousands of miles from the USA and any secure US communication/logistic lines? The answer is clearly, no.

Under any reasonable circumstances could the Soviets have expected the USA to fight over Afghanistan, an already communist state long in the Soviet camp and out of our sphere of influence? Certainly not after Vietnam, and probably not before. You and I know that and so did the Russians.

And as for the oil, the USSR is already the world's largest producer of oil and its future needs for middle eastern oil are debatable. But more importantly if oil seizure or an Iranian attack were their planned goals, what have the Reds gained from entering Afghanistan? Not much. The USSR itself already borders on Iran and there is a far better invasion route from the USSR itself right into Teheran than via Afghanistan, through some of the toughest terrain in the world for war fighting.

But if all of this makes you feel better, don't. Although I am saying that the USSR did not act due to US weaknesses and has no goal other than showing their own that communist governments cannot be overthrown by holy wars, I am also assuring you that their leadership will react savagely to perceived threats and use force against the weak when they are confident it will not lead to military confrontation with the US. That was true in the 1950s when the USA was clearly the strongest nation of all, and just as true or even truer today when the changed military facts of life prevent any one nation from being as singularly superior as the USA was then. Not a pretty world is it? Plop, Plop, Fizz, Fizz.

Red Propaganda Backfires
On Them and Us

8/16/81

"I do not believe that ———— the government tells us. What really happened?" Those are the words of a Russian policeman who at 2 a.m. was asking me what really happened to cause the death of one of their cosmonauts. Russian people are so accustomed to hearing lies from their government, they have learned not to trust government statements about anything. Lenin believed that information control was crucial to the revolution and immediately after his takeover developed an apparatus to control all forms of the media.

But control was not enough. Lenin also wanted to ensure that all news was coordinated and directed toward enhancing the goals of the revolution, even if it took lies to do so. Thus, the honorable profession of agitator and propagandist (Agitprop) was created and flourishes to this day. Nothing is legally printed in the USSR without some shaping and approval from the Agitprop organization of censors and creative PR.

As a result the Soviet Union is awash with banners, posters, signs and slogans. To the outsider it is somewhat overwhelming. American visitors seeing so many red and white signs and banners usually pick out a few of the most repetitious ones and nickname them. "KPSS," for example, the abbreviations for the Communist party, appear almost everywhere and are called "Howard Johnson" signs by many Americans. Other common slogans are named for McDonald's, Burma Shave, etc. They are advertising, of course, but for one product only—the continued control of the single party of the USSR.

Soviet citizens tend to take all of these signs for granted, seldom seeming to pay any attention to them, except when they are being taken down. When any such big replacement program begins, Soviets become uneasy and worry, wondering if the replacements will signal some change in party policy. They have to wonder if what they have been saying and doing, as of late, always planned to meet the current party line, will suddenly turn out to have been the incorrect viewpoint. They usually breathe a sigh of relief when only refurbished signs go back up without reflecting any change in leadership or path.

One of my favorite slogans is the one that says, "Communism is the future light of the world." My fondest recollection of that homily was seeing it freshly painted over the entrance to Polyclinic #48 in Minsk (I think I remember the number and place correctly). The clinic was in a new neighborhood of tens of high rise apartments, usually referred to as "instant slum" because of the shoddy construction; shabby, if any, landscaping; and dirt roads and paths. A young woman with babe in arms was exiting the clinic and stepping off the steps under the sign into about a foot deep mud puddle. To me the scene epitomized the USSR with its loudly proclaimed aspirations contrasted by its rather inferior achievements.

One of the USSR's top comedians was joking about the agitprop and censorship program one evening during the grand celebration of 50 years of Soviet power. He paused in his routine to compliment the Russians about their great societal achievements. He quoted statistics citing how many books are read in the Soviet Union annually (the figure included programs for the theater, ballet, opera, circus, etc.), how many people attend cultural events as compared with citizens of selected other nations, etc. He concluded with a comment about how many more Russians than Italians attend movies. But then he paused, winked at the audience and said, "Of course if we had movies like they do in Italy, we would attend even more." The audience broke up with the laughter of conspiratorial agreement.

This Soviet penchant for misspeak does them far more harm than the officials seem to realize—it creates an air of mistrust and suspicion that permeates every aspect of Soviet society.

As an example, Aeroflot, the Soviet airline, has an excellent safety record, primarily because of an extremely cautious operation made easier by the lack of competition. But Russians do not believe the safety claims since the government tries to cover up even the smallest of accidents. As a result the rumor mill runs rampant and most Soviets consider the line a risky endeavor. It is carried to such a stupid extreme that even minor mechanical difficulties are instead labeled as weather caused delays. As if the passenger cannot look outside and see the sun and the mechanics working feverishly around the plane—but communist engines can't malfunction!

This lie syndrome does have impact on those of us outside the USSR as well. Foolishly we often use their data to justify our own needs, say, in the defense area. One or two American defense analysts, for example, have built their careers out of touting the success of Soviet civil defense (CD), and based on Soviet data they posit it will give the Soviets an edge in nuclear war. Russians who have participated in the sham of a CD program, belie the claims and laugh at the exaggerated Red propaganda about its effectiveness.

Like so much of communist Russia their CD is an agitprop success but a hollow house of matches in reality. But it is one that can cost you and me "mucho dinero."

10/26/80

The Lie Syndrome

Soviet society is quite alien to the frame of reference of most Americans and the "lie syndrome" is one of the unique features which must be understood in order to fathom the USSR.

What do I mean by "lie syndrome"? One needs only to ask almost any random Soviet citizen and the answer will be that the government lies about almost everything, and since in the USSR the government is the owner/operator of all enterprise, almost every statement made and every apparent reality is either a lie or built on lies.

Let me illustrate with a true story. The director of a Soviet institute tasked with building large generators for high priority defense and research institutes calls his executive staff together and poses a not very unusual problem. They are due to deliver a giant generator to a scientific research center over a thousand miles away, but the generator is not ready and the necessary completion materials are not available. The director has tried to acquire the needed resources on the black market and from expeditors that operate outside of the complex national plan system but to no avail. What to do?

One of the bright young executives suggests they send a message to the scientists informing them that the equipment is ready and has been shipped, asking them to notify the senders when it arrives so that they can then dispatch an installation team. This lie, he suggests, will get them off the hook for several weeks before the scientists begin wondering why the generator has not arrived. When they finally do ask, the builders can blame the delay on the transportation service, which everyone knows is in a mess and loses half or so of what it handles. This is another lie itself but one widely believed. These two strategems should provide sufficient time to acquire the needed materials, build the generator and finally ship it as a substitute for the "lost one." Conditioned to such lies the group concurs and the message is drafted and the scheme launched.

The scenario looks good until only a few weeks later, with the generator still not ready, they receive a message from the scientific institute saying, "Generator arrived, send installation team." Lies begat lies, it's Catch 22, Soviet style!

This is not an isolated incident but a near routine occurrence in the USSR today. Still, many US Soviet specialists judge the status of Soviet war preparedness, production output, average income, or whatever based primarily on data published by official Soviet sources.

The result is a misled West and a USSR where only the innocent believe what their government says, and where most trust only in what they see themselves or can confirm from a trusted friend. Knowing this, the government finds itself telling more and more and bigger and bigger lies trying to con the populace into not believing that the previous lies were lies. Remind you of Watergate?

The lie syndrome thus contributes significantly to the inferiority complex the Russian leadership so often exhibits in its dealings with the rest of the world and its own people. It has also contributed in no small degree to the now almost total disillusionment of Soviets with the Communist dream. Twenty-five years ago it was not difficult in the USSR to find people who believed in Lenin's prophesy even after the evils of a Stalin, whom they labeled an aberration. Many then truly believed in the myth that communism would provide a paradise for the working man and serve all of the people better than other economic or political systems. They not only believed but were willing to work long hours and even sacrifice their lives for the good of the cause.

But today even in this birthplace of communism, true believers are few and far between, whether one looks at the party, the military, among the working class, or the intellectuals. Lenin's system has not delivered and at last, even the Russians are no longer kidding themselves about it.

The lies in the Soviet Union began under the guise of necessary untruths required to ensure the survival of the party during times of perceived extreme threat, They were judged acceptable because the "end" was thought to be glorious even if the "means" might be temporarily immoral. But the lies got easier and easier to speak and justify each time, until today no one is quite sure of the difference between reality and propaganda.

9/11/83

The Tragedy of 007

The Soviet destruction of the Korean airliner was tragic, stupid, and irresponsible. But is it proof that the USSR is the most vile and evil of nation states and unworthy of responsible relations with the civilized world, as many are now suggesting? Of course not!

Despicable as the Soviet action was it differs not a whit in its arrogant use of power from the recent Israeli destruction of the Iraqi nuclear facility, an unprovoked attack on another sovereign state's property and people. But the US continues civil relations with Israel and does not consider it the focus of all evil, even after they followed that atrocious act with the invasion of Lebanon, the bombing of Beirut, and the sickening refugee camp massacres that resulted.

Logically, many might even consider the US secret bombings of Cambodia and our subsequent invasion of that theretofore peaceful land as an equally horrid example of the corruption that can come from excessive reliance on power.

And many other incidents come to mind: the Israeli kill of a Libyan airliner that strayed over the Sinai about ten years ago (Golda Meir at least announced regret); the French violations of humanitarian ideals in Algeria; the British massacres of Indians during India's independence struggle; the similar and frequent US atrocities against our Native American populations; and our overreaction in the Mayaguez incident and the resultant deaths on both sides AFTER THE US PRISONERS HAD BEEN RELEASED.

All of the above is not said to excuse the Soviets for their despicable act, but to place it in a proper perspective. All of the incidents noted above were wrong, but none of them were committed by people who either thought themselves evil or who were bereft of redeeming values.

In fact I suggest they were all caused mostly by a climate that has dominated this globe for most of human existence. A condition of irrationality that results from fear, the fear of the unknown and a false belief that only you and your kind are really good, and thus anyone who thinks, looks, or acts differently is obviously inferior and probably evil. Korean/Soviet relations are about at that plane.

The current rhetoric, primarily by politicians, about the Soviet shootdown serves no valid purpose other than to enhance that climate of self-righteousness, which so contributed to the act in the first place. For this Soviet murder can only be explained by fear.

The Russians ARE paranoid, and so are most of us that live in this world of nuclear weapons and economic and religious rivalries. That paranoia pollutes our globe with apprehensions so powerful they can cloud our reasoning. We scream and holler about Soviet intrusions into our backyard, as if a Soviet brigade in Cuba can really threaten us; yet we surround the USSR with listening posts, a ring of bomber bases, and we and our allies outspend them on weaponry by billions, all the while arguing that our acts, unlike the Reds', only serve to make the world safer. And the Soviets think the same.

The truth is we both scare the "bejessus" out of each other. The major failure is that both sides are too eager to seek retaliations for the other's acts and cast the first stone, when what we need is more talk, visitation, efforts to understand, and reduced rhetoric.

But that is difficult. It is far easier to castigate, to point accusingly at the other and to beat our breasts in anger and indignation. We somehow fail to see that such only pollutes the atmosphere more and perpetuates the actions that to date have failed to make it a safer or saner world.

There have been two recent intrusions into Soviet territory. The first had a happy ending when the Greenpeace members, arrested for illegally entering Soviet territory, were treated well and quickly released. The second occasion occurred in a more sensitive area, involved a nation with unhealthy relations with the USSR, and was a repeat offense. The Greenpeace arrest showed us the Reds can be civilized while the airliner fiasco raises doubts. But let's not get carried away by our reactionary instinct to prove our humanist superiority, and instead concentrate on things that might make a recurrence less likely.

5/15/83

"Pssst Buddy, Wanna Start A War?"

The biggest surprize I discovered during my recent trip to the USSR with the US/Soviet Citizens' Dialog was the pervasive presence of a fear among Soviet citizenry that Ronald Reagan believes he can fight and win a nuclear war. They pleaded with me and my associates to bring the message home that they disagree and want peace. Such comments were made over and over again by people from all walks of life, little old ladies coming out of a church on their Palm Sunday, a Kazakh pilot on his way to the men's room, members of the prestigious US/Canada Institute, and many many others.

We had hardly left the airplane in Leningrad before being whisked to that city's new war memorial to the 900 days siege in WW II. The Soviets lost 20 million in that major war and they have gone out of their way to ensure that their populace remembers it. Every town features a war memorial of some kind where visitors and Soviets alike are reminded of the devastation WW II wrought on their land.

I had seen many such places before when I worked for two years at the US Embassy in Moscow, and I had expected that some of that emphasis on WW II might have died out in the intervening fifteen years but I was wrong. If anything it has been resparked by the growing Soviet fear that the US has decided to seek nuclear superiority. I have often speculated that this Soviet obsession with the so-called Great Patriotic War was hypocritical. Indeed, is it not true that it was Stalin's immoral act of allegiance with Hitler that allowed that monster to begin his efforts to conquer the world? And did not Stalin kill more of his own than the Soviets lost to the Germans?

But I must confess that this anti-war campaign, and it can be labeled as nothing else, serves a very useful purpose of not only glamorizing the heroes who fought for communism, but for making the Soviet populace the most aware on this globe of the horrors of war and thus ripe for peace movements, and that is not all bad. One may occasionally still hear an American say, stupid as the reality of it may be, "what we need is a good war," to either revive the economy, regain our superior status, straighten out the youth or whatever, but such words are not uttered by Soviets for they would be looked upon as blasphemous.

So I was not as startled by the Soviet obsession with the war and its reality as I was by the newly discovered fear that the US wants a war. That did not exist when I was there in the 60s but does today, and unfortunately our leadership has supplied the ammunition for its growth. Now some might be happy to hear that the Russians fear us. I guess even I could be if, as the proponents of American superiority suggest, the result would be a Soviet willingness to get serious about arms control. But the result appears to be just the opposite. The Soviets instead are reacting just as the Reagan administration wants Americans to react to their mythical tales of Soviet superiority, by supporting increases in their military arsenal. Here the result is the MX, cruise, B-1, Stealth and the more than fifteen per cent increase in US defense spending. And those results will be matched by the Soviets who are simultaneously winning the propaganda battle in their homeland and in much of the world as they accompany their build up with responses like, "the no first use", and nuclear freeze proposals that our leaders reject, leaving us looking like the arms race generators and them as only the responders.

To clear up that confusing and incorrect depiction of America's nuclear war attitude requires new initiatives, thus the birth of the US/Soviet Citizens' Dialog. This group's goal is to create a dialog between responsible citizens of the two superpowers to help clear up some of the obfuscation, misunderstandings and fears. We have brought Soviet citizens into US cities and homes and asked them to meet with average Americans and realize what kind of a society we have and what our goals are, while allowing Americans to do the same in the USSR. We dialoged together about some of these problems over the last few weeks and found an amazing amount of common ground, even while acknowledging some very basic differences. I want to talk about that trip in the next few columns and hope you will find my findings interesting and encouraging. But this first column, of necessity, had to deal with my most alarming discovery, "fear." We should be taking steps to reduce the fear but unfortunately our current policies seem to be doing just the opposite and that bodes poorly for all of us. But I believe when we know more about that other nation, perhaps we can devise more productive approaches.

5/29/83   # All Bad It Ain't

After a 15 year absence, the most impressive change I noted on a recent trip to the USSR was the improvement in living standard. The clothing, food, autos, department store stocks, and almost everything one can observe indicates that life is more rewarding materially today than it was in the late 60s. Of course that should be true, but I was still surprised by the degree of improvement.

I had been reading the Western media and many Western studies of the economic problems of the USSR and was lulled into expecting only minimal improvement. Thus I was impressed by the variety and quality of clothing, shoes, apartment conditions, etc. Of course all of this is true only by comparison to WHAT WAS in the USSR, not WHAT IS in the West.

In the late 60s when I went to an evening performance of the ballet, opera, theater or whatever, be it in Moscow or in the hinterlands, any Westerner had to take notice of the lack of variety of clothing models and styles. To put it bluntly one usually saw about three dresses and an equal number of shoe styles, both, more often than not, ill fitting and shabby in appearance. Each evening out this trip brought an entirely different circumstance. I almost never saw the same dress, the colors and varieties of clothing and shoes being wide and attractive. Now to those of us in the USA, the theater may sound like a phony place to make evaluations about general living standards, but in the USSR that is not so. Ordinary people can afford to go to the ballet and opera in the USSR, it's really quite inexpensive, and they do so. Working class folk often go right from work to the theater, eat their evening meal there, and then go home.

These observations lead me to a sad but important conclusion. We in America have a rather outdated and frozen view of the USSR furnished to us by our media and by our own inclination to lock the USSR into an image implanted in us in the 40s and 50s, Stalin's dark days of near poverty and an all-pervasive fear in the USSR.

That picture of the Soviet Union is no more valid today than the picture many Soviets carry of us, the Capitalist society depicted by Lenin and Marx, with enormous numbers of poor and unemployed, the soup lines, and child labor operating 12-hour days in the sweat shops for 25 cents a day. That is why I consider it so important for Americans to visit the USSR and vice versa, the goal simply is to get as many informed people as possible about life in the other superpower's domain. Soviets who visit the USA are invariably surprised about much of what they see and hear and the same for our visitors to their land. Each sees some things they do not like, but more often see many things that pleasantly surprise them. For instance, I still find it remarkable that with all of the vitriolic diatribe going on between our governments, most Soviet citizens are real admirers of America. They fondly remember our alliance during WW II and the Lend Lease aid.

How is this erroneous picture of one another developed and perpetuated? A recent PBS examination of the US media coverage of the USSR, which it concluded was almost criminal, suggests the answer. The media of each side accentuates the negative in their coverage. Any Soviet TV crew or reporter can come to the USA and bring home a factual, but still misleading story of our shortcomings, by simply haunting the urban slums or the Appalachia region and reporting it as the norm. They can choose to ignore the fact that the vast majority of America is vibrant, industrious, productive and providing the highest living standard and most freedoms in the world.

Similarly Western reporters can, and unfortunately most often do, report almost only on the abberations instead of the norm. Partly because some of them seem not to like their jobs, can't speak Russian and have little understanding of the USSR, and partly because the Soviets treat them poorly and differently than in the West and thus breed a contempt for the Red system.

But this ignores the reality that most Soviets are making the system work for them, improving their living standard, getting educated better than ever before, and seeing better opportunities for their children than they ever could have imagined for themselves. Certainly they have far less to say and influence on their govenors than we do. And they have drastically limited freedoms compared to ours, but that was true before Russia became the USSR and many a Soviet can and does argue effectively that they do have, through their Soviets and participation in volunteer groups, more influence than Russian Empire citizenry ever had before.

They cite, for example, the fact that they have gone through four constitutions in the last sixty years, all of which have been modified as a result of citizen discussion, and that in the last 25 years they have even recalled some 12 delegates to the Supreme Soviet and hundreds from the lower level legislatures. Additionally, in every election they reject over a hundred of the Party's choices for public office at one level or another. The latest data indicates more than 200 rejections in 1981. Their influence over the real rulers, the Party and the "nomemklatura" is truly less but does occur.

Life in the USSR is not a bowl of cherries, does not approach the worker's paradise Lenin promised, and certainly falls short of what most of us in the West would require to meet our minimal standards of acceptability, but in most Russians' view that is judgement by an inappropriate scale. Life is better than it was, more satisfying, more promising, more fulfilling than it was even twenty years ago. Many admitted to me their disappointments but most quickly added, "it is getting better, and when viewed by our different measurement, we are not only making progress but our system already offers some advantages that even your vaunted system can not match." They would then cite total employment, total health care, 30 day vacations for all, bilingual education, enhanced access to cultural events and entertainment, and other benefits which may not seem like much to us in the West but are marvelous achievements for the former Russian Empire.

All bad it ain't!

6/5/83 # True Confessions

All right, I confess, my recent trip to the USSR confirmed what the president, Richard Pipes, and conventional wisdom have been telling us: the Soviet Union is an awful place, the cancer that is infecting the world with instability, the focus of most evil on this globe. Additionally, its economic system is a mess, all we have to do is put a little more pressure on and it will collapse. The Soviet people look and act like drones as they wearily go back and forth to their dull work, getting drunk whenever they get a chance in order to escape the misery of communism. Religion is non-existent, the jails are overrun with innocent people who only wanted to speak the truth, and their military is ten feet tall and growing, determined soon to launch the final offensive that will enslave the world.

I wish I could write such words in honesty for it would certainly take the heat off. My columns never receive such a barrage of critical letters and phone calls as when I buck the conventional image and try to report objectively, as I see it, on the USSR. It seems that many Americans, thank goodness I do not think it is most, have room only for the worst about the Reds, they are not interested in the truth but only in reconfirming their preconceived notions that the USSR is the nastiest abomination ever to befall our planet. Sydney Harris said it well many moons ago when he reported that people prefer to be blissfully happy in their incorrect assumptions rather than to be confronted by truths they do not like.

One of the great mysteries to me is why almost any time I say something not severely critical about the USSR, I get roundly criticized for being anti-US, even if I have not mentioned the USA at all. Additionally if I criticize something the USA is doing , the callers and writers jump on me for being pro-Russian, asking me why I did not include in the column revelations about how nasty the Reds are, even though the column was not about the USSR at all. I have concluded they believe that America has to do some bad things, but they are done only to protect us from that horrible evil communism, and thus forgiveable.

Such alleged patriots are making a great prophet out of former President Eisenhower who warned us that the greatest danger he saw from communism was not that it would take us over but that we would become like the communists in the name of protecting ourselves from them. I must confess I share his fear, especially after the barrage arrives, when I have had the nerve to say something like, "although communism has proven itself a failure as far as a competitor with either capitalism or socialism in providing a standard of living, it has improved the economic conditions of people in the USSR and elsewhere."

But that is the evidence, confirmed by the most extensive CIA studies and by the less biased US Library of Congress, and for my part recently verified by a return visit some 15 years after serving at the US Embassy as a military attache. Of course the Soviet Union has flaws, primarily in the area of human rights deprivation but also in the efficiency and effectiveness of its economic machinery, but much of that is the result not of the "ism" itself but because its leaders came out of a culture where power was almost always concentrated in the hands of a few and that outcome was expected by both the leaders and followers of this new way. By perpetuating, indeed even to some considerable degree perfecting that ideal, even in their revolutionary new approach to life, they have assured their failure to provide human rights commensurate with their other progress. But believe it or not and like it or not, many a Russian and other Soviet prefer the stability that brings to what they would view as our greater freedom but concomitant chaos.

It's about time we quit trying to work with myths that needlessly make us feel morally superior and begin dealing with real facts. Facts such as those revealed in Andrew Cockburn's new book on the Soviet military wherein he shows that they have more weaknesses than strengths. Or as Harold Berman, the Harvard expert on Soviet law, tells us in *Newsweek* that the Soviet system offers its followers economic security and power, and that even its legality is improving with the times. Facts like those repeatedly drummed into us by George Kennan, perhaps our most experienced sovietoligist, who has concluded that the USSR of today is not the same nation of Stalin's time and must be dealt with by different strokes, strokes that recognize and respond to their positive changes rather than just drawing polemic lines in our relations and concluding that our styles are totally incompatible. Or as John Kennedy said in perhaps his most eloquent speech, "No government or social system is so evil that its people must be considered as lacking in virtue."

There are a lot of nations in this world that Americans, with their justly noble ideas, have every reason to dislike, but we often overlook their shortcomings in the name of practical choice of the lesser of two evils, preferring anti-communism to most other alternatives. But I suggest a more realistic approach would be to start judging the USSR without our red blinders. In so doing we may find much of the ugliness, although truly not all, was long ago displaced by a leadership that today is more stodgy bureaucracy striving for stability and a measure of economic success, than revolutionary Mafia out to blow away the forces of good who might hinder their profit.

6/12/83Good News, Bad News

So what is the good news, bad news on the USSR? The bad news is evident; it is still basically a police state ruled by a small number of elite who can and do act as they please almost without constraint; a fact that is clearly shameful but also not unusual in the vast majority of the nations on this globe.

But there is also good news and I can report on several areas where my Soviet acquaintances brag that their life is as good or better than mine.

The biggest brag I hear from Soviets is over their employment status versus that in the West. A virtual (almost but not quite) "total employment is one of the real benefits of our system over yours," they posit often. " Why we can not imagine 10% unemployment and cannot understand why in a so called democracy it is tolerated." The bad news of this issue, of course, is that there is as much underemployment, or more, in the USSR as unemployment here. Soviet citizens are assigned their first job by bureaucratic fiat and it often is well below the capability of the employee. He does, however, have a choice. In spite of a strong belief to the contrary here in the West, Soviets can and do change jobs although the process can be discouraging and often requires a bribe or two.

Another brag is that their youth receive more opportunities for exposure to culture, i.e. the opera, ballet, classical music and literature, plus all kinds of opportunities via the Young Pioneer programs to develop their interest with the best of facilities and volunteer teachers. This is more true in the major cities, of course, than in the country side, but truly does exceed in magnitude anything we have in the USA from scouting to YMCAs.

Additionally, at the college level the USSR pays its students, those who have scored highest on national tests, to go for a higher education. It is that so-called gift from the nation, for which the graduate owes the state, that is used to explain why Soviet intellectuals are seldom allowed to emigrate from the USSR. On the negative side of this idea is the fact that an enormously high percentage of the national elite's children do well on the tests, even though they do not appear to be among the best and brightest—bribery again rears its ugly head as the answer.

Very inexpensive medical care and prescription medicine is another attribute Soviets gloat about when contrasting their life with ours. Soviet doctors will make house calls, and medicines, including some but certainly not all of the best, are available for a mere pittance. But as usual there is a negative side. Doctors in the USSR are not normally anywhere near as well educated as those in the West, they are poorly paid and often lack incentive. And as for hospital and medical facilities the best that can be said is that for the select few the facilities are excellent, while for most they are woefully outdated and often verge on the unsanitary if for no other reason than age. A group of US medical people flew out of the USSR on the plane with me recently, and they unanimously agreed that the USSR is 20 or more years behind the West's modern medicine.

Bilingual education is another area of Soviet braggadocio. In all of the national republics students have the choice of studying in their national language or in Russian. In either case the other language is also a required course.

Additionally, many Russian families can choose to send their children to special foreign language schools from the first grade on up. I visited an English language school where a friend's daughter was in attendance. The school, like most I have seen there, was lacking in facilities and under repair, but the children had a great attitude, and the teachers spoke excellent English. Such schools are only available in large cities but are quite popular along with other special schools for math emphasis, and of course, the famous athletic emphasis schools, all at no extra cost to the student who displays an aptitude.

We often hear a lot about the nationality problem in the USSR and it is probably true where one is talking about places like Estonia and Latvia that thrived as independent nations prior to being Sovietized. But for many of the republics, where their existence prior to annexation by the USSR was primitive, places like Kazkhstan, Turkmenia, Kirgizia and Tadjikstan, that question is more complex. Life for the vast majority of them has become quite modern, their women have changed from being pure chattel to humans with rights, education and a future, and when contrasted with their ethnic relatives in places like Afghanistan, Pakistan and even Iran; their living standard and future is clearly better and they know it.

There is good and bad news in the land of the Reds and we must be aware of both if we hope to deal with them realistically.

6/12/83   Transition Politics in the USSR

"And where were you when all of that was taking place?" Those are the words attributed to an undisclosed questioner who interrupted Khrushchev during his denunciation of Stalin in the Palace of Soviets many years ago. The probably apocryphal story goes on to say that Khrushchev interrupted his litany of Stalin's crimes, stared long and hard at the audience and with a stern voice asked, "Who said that?" The audience was hushed, with most too terrified to even raise their heads and meet the gaze of the great new leader. After the frozen silence had gone on for what seemed like an interminable time, Khrushchev smiled, nodded and said, "That's where I was too." He returned to the text. Brezhnev shares that legacy.

Sixty-five years after its creation the Soviet Union has lost its fourth major leader and the world still has little to mourn. For Leonid Brezhnev does have to share in the multifarious crimes of the great terrorist, Stalin, the man who killed more of his own people than any other leader in history. Still, like his immediate predecessor, Khrushchev, Brezhnev seems not to have borne that mantle proudly. Brezhnev's departure marks a coming of maturity for the USSR for he is the first of their masters to die in office of natural causes. Lenin was shot and never fully recovered, Stalin was likely poisoned by his closest followers who feared for their lives, and the wily Nikita was outsmarted by Brezhnev, the man he had brought into the leadership circle, and sent into inglorious exile.

Andropov's apparent quick succession to the throne also marks a change worth noting. Truly, the Soviet leadership deemed it important to show a stable face to the world even if the bickering and backfighting within the so called collective is, as is probable, alive and well. Their recognition that the world expects an important nation to act controlled and to provide a facade of unity under stress is also a sign that the USSR of today is not the militant aberration of a nation state that it was some sixty years ago—a myth that is unfortunately at the heart of US policy toward the new Russia.

But do not be fooled by the apparent tranquility. The succession question has probably not really been settled. Khrushchev only emerged as the leader to follow Stalin after other faces, now almost forgotten, appeared to be the heirs apparent. Andropov has taken a giant step in the right direction by being named Party Secretary but he still has battles to win. Brezhnev, Stalin, Khrushchev, and Lenin all held not only that job but also the post of Chairman of the Presidium before their power was reckoned as indisputable. And each of them probably held just a little less power than their predecessor at the peak of their reign, at least that is true since Stalin.

Most now expect that Chernenko will take over that other key spot and that we will see a real collective with the power dispersed between the above mentioned two plus Ustinov, who commands the military industrial complex and Tikhonov who commands the technocrats. Lurking in the wings with a real shot at becoming number one someday are Grishin, Romanov, Gorbachev and other men about whom we have heard very little.

And what does all of this mean for the USA and the rest of the world? Well obviously no one can say for sure, but based on Soviet history and a lot of observation of the current scene it should bode well. The Soviets are clearly in for a time of reassessment and concentration on the internal scene. This is not a time for being pushy with them since the new leaders might feel they are being tested and have to respond aggressively to prove their ability and courage. But it is a good time to seek to create a dialog for they will each want to show that they can get better recognition in this world for their system, and that they can reap more benefits than those they followed into office.

Brezhnev was clearly not a man for whom the doors of heaven will be thrust wide. But as evaluated by men and women who judge his rule by Soviet or even historical Russian rules, he fares rather well. The lot of the Soviet peoples has not even approached the paradise Lenin predicted or the goals promised by Khrushchev, but life has been far better under Brezhnev than under any of those earlier leaders in both the economic and human rights areas. The rules of life are now more defined, the fear has been lessened, and the pocket book and stomach swelled. Brezhnev contributed significantly to the concept of detente and will be most remembered by Soviet historians for his emphasis on peaceful coexistence, the avoidance of a nuclear war, and his efforts to improve relations with the rest of Europe. Many a European and Jewish family has been reunited thanks to his policies. If Andropov does as well the world will benefit.

George Kennan keeps trying to tell us that the USSR of today is not the same nation for which he designed the containment policy of some thirty years ago. It's not a nice place either, but a better one than it was and with a shot at becoming even better yet; thanks to the atmosphere created by Mr. Brezhnev. The new Soviet leadership is at a crossroad in Soviet development and they may choose any of several alternative courses. We should do everything possible to convince them that the way Brezhnev pursued in the heyday of detente is the one most likely to offer them continued rule and increased prosperity. To do so will require a change in the policy we have pursued toward them in the last two years. We are a far more flexible state than they, I hope we will prove it in the crucial weeks to come.

To Those "Seewhatyouwanttoseeonly types"

10/2/83

A few of my readers, whose eyes are apparently impaired by a not so rare disease called "seewhatyouwanttoseeonly" have concluded I am a Soviet/communist sympathizer and apologist. A few quotes from my columns over the the last couple of years are repeated here to help them learn to overcome their apparent myopia.

- The Soviet leadership is best compared with the Mafia.
- The communism of Lenin and Stalin is flawed and has committed some awful sins.
- Almost every leader in the USSR has been so corrupted by excessive power that he might look clean only by comparison with his peers, while in reality he is totally contaminated.
- The "Lie Syndrome" is one of the unique features of the USSR.
- All communist states reveal horror stories of freedom deprivation, torture, censorship and despotic rule.
- Communism is simply not compatible with freedom. . . .
- The real losers . . . will be the Soviets, whose use of force simply affirms the emptiness of their system.
- The Soviet party simply cannot let such freedoms exist and survive in power.

Since these are only some of the many condemnations I have made of the Reds, it must be my other comments that upset the critics. Words that suggest the USSR, bad as it may be, is not all that unique in either world history or contemporary times.

By the highest accounts there are now 165 nation states on this globe; yet, few will argue that more than 25-30 of them can be credited generally with the routine practice of human rights toward their citizens and others, with letting their citizens be truly free, and with providing materially for the majority of their people. Additionally most have poor records when it comes to accepting the rule of international law over their national interests. But this has not prevented the USA from having good relations with more than 125 nations and from giving strong support to many of the worst offenders. Thus if the USSR is an outlaw nation it can still consider itself in the majority and wonder why the US singles it out for opprobrium.

Although I have never suggested that basic freedoms exist in the USSR, I have argued that their living standard and the guarantee of work, vacation, health care, and education indicate that communism has been more successful than its detractors like to admit. The fact that more people in the USSR are now living at a higher standard then under any preceding system is one of the major reasons there is no incipient revolt just under the surface in the USSR. People who have done without the basics for a long time will be satisfied with just them for awhile before they begin demanding more. The Soviet people do not have those freedoms but a slowly growing number of Soviets are finding more and more ways to gain them even within a system that resists.

My columns on the USSR also often suggest that we are more likely to help Soviet citizens acquire some of those basic freedoms by increasing our contacts, trade, discussions and quiet but firm diplomacy, than by accusations, threats, the arms race, and a propagandistic dialog that indicates they are vile and cannot change.

The facts are the USSR has changed and almost exclusively in the direction we would like to see. It is clearly a better place to live now, materially, spiritually, and even freedom-wise than it was in the '30s through the '50s by any standard of measurement, a fact that is even confirmed by CIA studies and which I have personally verified.

By a less confrontational policy I do not suggest that the USA should give in to every Soviet demand, ignore their violations of civilized norms, or disarm unilaterally in some vain hope they might follow. But we can talk and deal with them as we do with other nations, expecting them to react, not in our interest but in theirs, and when those interests are mutual, as they clearly are in limiting the arms race for example, to negotiate agreements and adhere to them.

Nuclear arms reductions or even a freeze are particularly in Soviet interests since it takes twice as high a percentage of their GNP to just match our expenditures, and since such agreements make them look like our equal even though the US is in reality a far stronger overall world power.

<table>
<tr><td>3/17/85</td><td># Dealing With Reality</td></tr>
</table>

"Why are you so critical of the USA, and why do you defend the Soviet Union?" Questions similar to that are so frequently asked of me that I feel a column on the topic is warranted.

I do not enjoy being critical of our nation, and only do so in the spirit of patriotism. I want our country not to be just good, but as good as it can be. In order to help it achieve that loftier goal, I hold it to a high standard, an even higher one than I set for other nations, and I feel compelled to speak out when I perceive it falls short of such an achievement. I consider the true patriot not to be one who believes "My country right or wrong, love it or leave it," for that is like the blind adoration of an ass for its stall, manure infested and all. To me the real patriot is the one who observes his nation's every act with a critical eye and asks if it measures up to the dream of Jefferson et al.

As for the USSR I am neither an admirer nor a blind automaton mouthing the popular cries of "shame." I do not hold that nation to as high a standard as I do the USA. Partly because it is a much newer nation state risen out of violent revolution, and held back by far more forces that have made its final shape harder to mold. But additionally, by the fact that I have studied it intensively, and know that it was shaped historically by cultures and events quite different from those in our past, and that it, as all other nations, inevitably must be judged based on its, not our, past.

All that does not allow me to ignore its failings, only to better understand them. I deplore their lack of freedom, and the overconcentration of power that harms so many both internally and externally, but I can also see some progress, slow and incomplete as it has been.

Americans have a tendency to expect every nation to be judged by our current standards immediately after its creation. They ignore the reality that it took us decades to free our slaves, turn from jailing those who were critical of the government, allow non-property owners, women and minorities to vote, and to treat our minorities with even near equality, even though our constitution demanded it from the start.

I neither admire the USSR nor hate it, but I do try to describe it as it really is, flawed of course, less than it should and could be, but not as rotten and without redeeming value as some insist that it is.

But I suggest it is the others, the repeaters of the conventional wisdom, who dangerously tend to mislead. For they create a false picture of implacable foes, armed to the teeth, no hope for any compromise or room for understanding and coexistence. They insist we spend our valuable coin on ever more armaments to either gain or maintain some ethereal superiority, or suffer ignoble defeat at the hands of the most dangerous vermin that have ever inhabited this globe. They create in us, as do their counterpart Soviets, and there are many of these folk in the USSR as well, a sense of rivalry, hatred, and danger that spurs a madness. An insanity that is pushing the entire globe closer and closer to an Armageddon. An Armageddon that seems not only more likely yearly, but also more acceptable because, horrible as it may be, it will also rid this planet of those atrocious Reds or Capitalists.

But Kennan, Draper, Zuckerman, Bethe, Feld, Ball and so many many others, who have met the Soviets up close and who understand modern weaponry, know that the myth and the reality of the USSR are different, and that in the name of sanity we must pause in the arms race and start thinking rationally about our rivals. Men as evil as they are portrayed could not have built the society they have, achieved their scientific, technological, and even humane successes in finding jobs for all, medical treatment for the masses, art, music, education, etc. If they were evil incarnate, long ago they would have attacked many more, poisoned the earth and the skies and still be doing as despicable deeds today as Stalin did anon.

In my mind better informed men exhort us to recognize that attempting to change the Soviets by bellicose threat and military intimidation have merely

brought the globe closer to extinction. A better way would be by quiet example while seeking to find a common ground of interest in trade, arms control,
environmental protection, scientific progress, joint efforts to save the have-
nots, and a firm insistence on reasonable behavior.

But that is so hard to do when hatred and fear prevail, even if the hatred is
irrational and our fears of our own making. Soviets really are people and they
can and must be dealt with as such, or else we shall perish together.

CHAPTER TWO

The Arms Race

One of the things I most want to change about my fellow Americans is their tendency to avoid thinking about the arms race. Far too many either fear this issue so much they actively try not to think about it and hope it will go away, or just escape it by concluding that it is so complex a subject only our leaders have the time and access to adequate knowledge to make valid decisions about the issues.

I am convinced otherwise, convinced that the nuclear arms race is far too important to be left to political leaders, scientists and generals alone to decide, and that reasonably informed citizens can readily understand the subject and make valid judgements about it. I am also convinced that democracy works well only with a well informed electorate, and that it is therefore our duty to become familiar with the most important issues of our day. Clearly the nuclear arms race is one such issue.

Thus I often wrote about the arms race, usually critically of US policies, in opposition to proposed new weapons systems such as the MX, Euromissiles, and Star Wars, and in favor of arms talks and in particular a US/Soviet mutual nuclear freeze.

But in spite of those feelings I am not an advocate of total nuclear disarmament. On the contrary I think that is an impractical, dangerous, and foolhardy idea. The nuclear genie is out of its bottle and cannot be stuffed back in and forgotten, as idealistic a prospect as that may be. I believe we have to learn to live with nukes, with their good and bad, and as so often is the case there are both benefits and dangers in this invention.

The knowledge of nuclear weaponry is now commonplace. Thus, even in the highly unlikely chance that all of the nuclear powers could agree on their danger and a verifiable means for eliminating them, I believe they would be foolish to do so. Because somewhere some nation or group would see in that disarmament an opportunity for them to acquire nukes and power over those of us who had disarmed. It is the first duty of every government to provide its people the best security within reason, and today the only defense against domination by those with nukes is nukes. One of the benefits of the nuclear invention is that it provides the nearest thing we have known in history to a perfect defense, to date no nuclear possessor's borders have been seriously attacked by any aggressor. Nukes also bring political clout that is clearly beneficial to a nation.

But believing that nukes will be with us for the duration does not prevent one from opposing their almost unlimited growth in numbers, size, accuracy, ease of delivery, firepower, etc. For it is also my firm belief from virtually years of research and work in the arena that nuclear weapons are the most unique weapon ever developed and that they cannot be treated as other weapon systems.

What is unique about nukes? First, unlike other weapons like tanks, airplanes, cannon, and divisions you cannot measure comparative nuclear arsenals by bean counts determining that he who has the most is most likely to be superior. Since nukes are the only weapons yet designed that have the capacity to destroy entire cities and complexes in seconds, most experts have concluded that after a certain number are acquired a nation has assured deterrence, and after that more are not needed to match another nation's foolish drive for superiority. Indeed more might even diminish a nation's deterrence.

That is the second unique point, since nukes have proliferated beyond one possessor and exponentially grown in firepower. They are no longer useful as an offensive weapon. They are only good for deterring war. But their irrational growth in numbers, accuracy, etc. tends to trivialize them and pushes experts to think of some possible way to use them and thus justify that expenditure and effort. But if used, nukes lose their positive value, which is deterrence, and become negatives, potentials for self-destruction and excessive overkill. Many an expert today scoffs at the idea of limited or even adequately controlled nuclear war, and I share their belief.

All of the above argues that the current arms race is not conducted for rational reasons. In the columns I attempt to explain why it is perpetuated and suggest how we can break the cycle and move to reasonable reductions that might make the likelihood of a nuclear winter less and our mutual survival in a nuclear age better. For a combined solution to the arms race, US/Soviet excessive rivalry, and the world hunger problem see the last chapter and a piece called THE SOVEUS SOLUTION.

Read these columns carefully, check out more info via the bibliography and seek answers for yourself. Then get involved. This issue is too vital to be left to others to decide. Our past policies have not made us or the world more secure. Perhaps it is time to consider something new.

1/1/84   Building A Doomsday Machine

Nineteen eighty-four is now here and it is up to us to prove that Orwell's ominous predictions about that year are wrong. In my opinion Mr. Orwell missed the boat when he predicted that governments and their thought control over their constituencies would be the most dangerous threat to humankind by 1984. He apparently did not foresee, as H. G. Wells did, that developments in warfare would proffer an even greater threat to humankind's survival.

But recent studies have suggested that the weapons we once thought could assure our safety may threaten the globe and its populace more than big brotherism. Speaking about such a study Carl Sagan says, "Conventional pre-1945 wisdom, no matter how deeply felt, is not an adequate guide in an age of apocalyptic weapons."

I have read several accounts of these research projects and believe the best and most informative for laymen is found in the Winter 1983/84 issue of FOREIGN AFFAIRS. Written by the noted scientist Carl Sagan, who was one of the major researchers on the project, the article suggests that, "We have, by slow and imperceptible steps, been constructing a Doomsday Machine." The "we" to which he refers is not the US alone but all of the nuclear powers, although he does assert that the US is the foremost nuclear power by virtue of its possession of more warheads than any other nation and its current plans to increase that warhead total significantly.

The article reports on two studies conducted by eminent scientists and confirmed by hundreds of their colleagues around the globe. The conclusions indicate that, bad as we always thought a nuclear war would be, we have underestimated its impact. They conclude there is a real danger that a nuclear war, even one using only about one-twentieth of the world's strategic nuclear arsenals, might well lead to the end of humankind as we know it.

The major new factor in their reasoning is that the fires created by a 1000 megaton war, requiring the firing of only a small portion of the forces available, would create a Nuclear Winter as the smoke, soot, etc. rises to the atmosphere and blots out the sun for months. The result would be a drop in earth temperatures that could prove catastrophic. They note that a one degree drop in the earth's temperature caused by a volcanic eruption in 1815 produced what was known in Europe and America as the "year without summer." But their studies show much greater drops inherent in almost any probable level of nuclear exchange; drops in temperature from 5 degrees to more than 50, and all leveling off eventually to a long term loss of 5 to 30 degrees. The impact on the food chain alone would be life threatening.

Naturally, there are other dangers noted besides the Nuclear Winter such as serious depletion of the ozone layer, the total inability of the world to deal with the more than a billion casualties, disease, and authority problems, plus the fact that the fallout will reach the Southern hemisphere after all. The report is devastating and as Sagan asserts demands that we rethink our strategies about nuclear war, its deterrent capability, and potential for use.

Sagan argues that since all the leaders of nuclear nations now know about these potential effects, "first strike" becomes a totally absurd idea, nothing more than suicide by the attacker, even if he succeeds in his goal of disarming the target nation. Sagan asserts the prospects for civil defense in nuclear war are negligible, and suggests some steps we could take that might diminish the threat. He evaluates a mutual nuclear freeze highly as a beginning move to make the globe safer.

I have long argued against the belief that nuclear weapons are just another means of acquiring military superiority. This new study strengthens my thinking. More is not better in the nuclear era, in fact, more only decreases our likelihood of surviving, while fewer make their use more credible, and thus their deterrent effectiveness more viable.

I urge each of you to consider as New Year's resolution number one in 1984, more important than giving up smoking, cutting out calories, or even avoiding a hangover on New Year's Day, to read the article in Foreign Affairs and then get active in the nuclear movement. Make sure that your nation adopts a sane policy about nuclear weaponry in 1984. It is your government, not Big Brother's, but its current policies emulate misspeak and violate every tenet suggested by these new studies. Let's show Orwell that we the people are too smart and too involved to allow governments to destroy us, either by mind control or nuclear madness.

Arms Race Madness

5/22/83

I am very impressed by the growing interest in the nuclear issue and my hat goes off to the Catholic bishops, the many Protestant churches, the League of Women Voters, and the numerous other groups that have decided to investigate the issues and take a stand on this most important subject of our times.

I have spent much of my adult life dealing with the subject and have reached several conclusions. While I agree with the bishops, for example, that nuclear war is unthinkable; I disagree with their hope for total nuclear disarmament. Not that I do not think that would be good— it's just that I do not think it is a plausible hope. The nuclear reality is now known by every nation on this globe and there is no way to assure that one nation or another might not seek nuclear domination if the others refused to defend themselves from that possibility. Nuclear weapons, like other not so nice items of reality, can be useful if used properly and awful if used improperly. Besides their necessity there are several other vital conclusions about nukes that the public must understand.

First, they are indeed a revolutionary weapon the likes of which have never before occurred in history. Second, they were only usable as an offensive weapon when only one possessor existed. Third, they are valuable to nations providing them the best defense against attack known to man and entry into a special international club with extraordinary influence on world affairs. Fourth, today they are usable only under the extreme circumstance of national emergency, when a nation feels as if it has its back to the wall and will lose all if it does not give them a try. Fifth, a nation can get all that it can gain from nuclear possession by possessing no more than a thousand deliverable warheads, lesser numbers may be just as good, and any more are a waste of money and resources, and even a danger for they tempt the ambitious to find a use for them and thus threaten humankind.

With all that in mind it is easy to see why I bemoan the fact that Congress seems poised to approve the MX missile after more than a decade of questioning its need in the US arsenal. Especially when that approval is coming at the very moment when the missile's justification has reached its lowest credibility. After Congress's last rejection of funds to deploy the missile, the President appointed a commission to look into the issue. Even though the commission was stacked with MX supporters its report, although recommending deployment, struck a terrible blow against the missile. It revealed that there really is no "window of vulnerability," the major justification used by the Reagan administration, and indeed for proof the commission recommended putting the MX in the very silos that were allegedly most vulnerable through that window.

They also recommended that as soon as possible our nation should develop and deploy a weapon that is the exact opposite of the MX, instead of big and multi-warheaded, a small and single headed, mobile monster. Thus indicating conclusively that the MX is the wrong missile for these times.

So here is a missile that the Air Force wants because it will give them a hard target kill capability but in insufficient numbers to be usefully employed. A missile that was to shut the window of vulnerability but does not, except that the window just disappeared. And this super weapon turns out to be merely an interim measure until we get what the Commission says we really need, a Midgetman. In other words, folks, we will spend about 20 billion and make no one happy except the MX manufacturers. A thousand more warheads will be added to our total of over 9000 even though almost everyone now admits that is about 7000 or so more than one really needs for deterrence.

But here it comes anyhow and apparently only because the president has written a letter to several key congressmen promising them that if they give him the MX, he will make some serious changes in his arms control policies. In other words the MX really is a bargaining chip after all but not with the Russians, a chip to be used between Congress and the president to finally get him serious about arms control.

I don't know how congressmen feel about the president's promises, but right now they do not sit very well with me. This is the man who four times now has signed his name to a letter to Congress certifying that human rights have improved in El Salvador. The snickers over that have reached a loud guffhaw, and he is the same man that just told Congress and all of us that the money being spent by the CIA on insurgents in Central America is not to overthrow the government of Nicaragua, which they have invaded, but just to interdict the flow of arms from there to El Salvador. The only trouble is he seems to have forgotten to put that note in the guerrilla paycheck for they profess just the opposite intent.

The MX passage may offer one of the worse case studies in American defense analysis. All the arguments say no, no but there is yes, yes in Congress's "ayes." We should all be ashamed.

3/24/85   # The Beat Goes On

If ever Americans needed proof that our Congress is not responsive to the wishes of the people, the latest MX vote should provide it. Roughly a year ago it looked like the MX was dead in both houses and could never survive the scheduled four votes, two in each chamber, needed for funding in '85. But for various reasons, and in spite of consistent polls showing a strong majority public opposition, the MX won support of 55 senators and looks like it will probably survive in the House as well.

Why? First and foremost because our representatives have succumbed to the administration contention that a vote against the MX would "pull the rug out from under the US negotiators at the newly started arms talks." Reagan, Shultz, and Weinberger have touted this view above all others, arguing that if Congress denies us this weapon the Soviets will conclude there is no need for them to compromise in the negotiations.

The weakness in that contention is that the exact opposite result occurred the last time an MX vote was held. The MX passed and shortly thereafter the Soviets, instead of making concessions, called the talks off for better than a year. Paul Nitze, an MX proponent and START negotiator, admitted the MX was not that important to the negotiations, but the myth prevails.

The second major pro-MX argument is that this missile is needed by the USA to counter a Soviet capability. Without the MX, the argument goes, only the USSR has missiles big enough and accurate enough to destroy so called "hard targets," targets like missiles in their silos. The US needs an MX to match that Soviet capability or be perceived as inferior.

The counter argument to that is that the US staunchly adheres to the position that we, unlike the Reds, would never start a nuclear war. But if we only intend to use our nukes as deterrents, in other words as retaliation to a first strike attempt by the other guy, we need no counterforce or hard target weapons. Such targets would mostly be empty holes by the time we launched a retaliatory raid. But even if the Reds did attack us first, and incredibly successfully, we would have 3000-5000 warheads left to retaliate with, the provocation of millions of American deaths, and the assurance that only a few hundred of those warheads can send the USSR back to caveman status. Only madmen would launch against such odds, and the MX changes that scenario not a whit. The MX neither gives nor takes away a first strike capability.

Argument three for the MX asserts that we need the MX because the president has asked for it. It is not good to show a divided national leadership. Americans elected this man with a strong majority knowing he wanted an MX and Congress should give it to him.

That argument is the weakest of all. Our constitution clearly identifies Congress as the place for raising and equipping our military forces. Once they do so the president is their commander, enjoined to use those forces as necessary to secure the security of our nation. A Congress that simply gave the president everything he asked for would be a useless rubber stamp a la the Supreme Soviet. Congress should act in the nation's interest as they see it, not at the president's whim. Polls clearly reveal that the public supports Reagan, like other presidents before him, while opposing many of his ideas, including much of his continued arms buildup.

The MX has a miserable history. The justification for it has changed every several years, and the commission created primarily to save it (the Scowcroft Commission) actually produced better arguments against it than for it. Stopping MX deployment would do more to enhance the atmosphere of the arms talks than building it can ever do for our security.

I am not an opponent of nuclear weapons. As terrible as they are, they are a reality and a useful one that has contributed much, perhaps almost all, to our nation's security during some dangerous times. But the MX will squander our coin, lessen the likelihood of reasonable arms control, and increase the likelihood and catastrophic effect, if that is possible, of a nuclear war. My hat goes off to those informed congressmen who had the guts and the intelligence to resist the pressures and the simplistic arguments and voted against the MX last week. We in Montana should be proud that both of our senators did so and our western district representative will as well. Thanks guys!!

4/3/83   # Master of Misspeak

The president's recent defense speech was a classic case of governmental misspeak delivered by a pro at making fiction seem real. There were many examples where I thought the president either misled or tried to conjure up panic without valid reason.

He attempted to pull the wool over our eyes when he recounted how the Soviets are building more of this or that kind of weapon in the last ten years. When a runner is behind in a race naturally he has to run a lot faster just to gain, but in doing so it does not guarantee he is going to win. Carefully the president avoided telling the American public what every newspaper in the country has revealed hundreds of times and what every military official knows full well, including, thank heavens, Soviet officials: THAT THE USA HAS ABOUT 2000 MORE STRATEGIC NUCLEAR WAR-HEADS THAN THE RUSSIANS AND THAT US MISSILES, EVEN IF OLDER BY CALENDAR MEASURE, ARE BETTER. In fact only the last of the five new Soviet strategic missile models the president noted is in the same league with its US counterpart.

Speaking of warheads the president did switch to warhead numbers when he got to an area in his speech where the Russians had more of them. In speaking about the intermediate range missiles in Europe he showed a chart indicating that the Soviets have some 1300 warheads to our none. It is true but how many of you noticed that the chart started out with the Soviets already at 600. That is because the Soviets have had this enormous lead in intermediate missiles for over a decade. Long ago we had deliberately decided not to deploy such missiles because they seemed extraneous to our needs. We already possessed enough strategic missiles to destroy the USSR, and also had a nuclear superiority in the European area as a result of aircraft, nuclear subs and aircraft carriers in the region. We still have them! But in the SALT talks we refused to include those weapon systems for negotiation so the natural reaction of the Russians was to get themselves a significant nuclear force that was also outside of the SALT limits. Unfortunately for them ours can still attack the USSR whereas theirs can only attack US forces in Europe.

But there is no need to match those Soviet forces, and in fact, if we do so we lessen the credibility of our support of Europe by making it appear we might be willing to fight a war limited to Europe.

The president, after defining deterrence as assured when the gains outweigh the risks, then suggested that the Russians were acquiring the ability to destroy all of America's land based missiles. But I suggest that defies his own definition. For if the Soviets did pull off such an attack, and that is a highly debatable possibility, the US would retain about 7000 nuclear warheads deployed in other strategic modes, and even Henry Kissinger was recently quoted as noting that a mere five hundred warheads are enough to deter any nation. The risks would clearly outweigh the gains.

Obviously, the president wanted something dramatic to make his case and he tried. Shades of the Cuban missile crisis: he declassified photos and showed us MIG'S and intelligence bases in Cuba, Soviet helicopters in Nicarauga, and a Soviet built runway in Grenada.

Are there no limits to Soviet audacity? The MIGs have been in Cuba for years and the US has hardly suffered, and we have surrounded the USSR with such intelligence bases for decades, but I guess only Russian intelligence collection is threatening. But I think he missed his score. Most everyone I have discussed it with so far agrees that for now the Texas Rangers could probably defeat the Grenadans, even with their new Soviet runway.

And finally the president reminded us of how the West's unpreparedness allowed Hitler to ravage. The reminder is nice but hardly relevant. In the '30s the US defense budget was miniscule and we truly were unarmed, but today the US is the world's largest nuclear warhead possessor, and the leader of an alliance that out mans and out defense spends the USSR and its allies. The US is strong and should remain so even if all we get out of Congress is that measly Democratic defense budget at over 200 billion bucks.

Finally the president asked us to be impressed with what his programs have already done for our defense. He noted that more people are joining the military voluntarily than ever before and staying in longer. Does he really think that has nothing to do with the over ten percent unemployment his policies have wrought, and does he now want to brag about it? America does need a strong defense and the Russians are a more formidable foe than they were ten years ago. But not through any fault of the current or past US governments. It is because the nature of military might has been enormously changed by the nuclear weapon. They have made deterrence of war mandatory and fighting wars mad. To me the president's budget priorities seem more aimed at fighting than deterring and thus should be modified. It seems obvious they will in spite of such shabby attempts to persuade the public otherwise. You are smarter than he thinks.

4/10/83   # Freeze Beats Alternatives

On March 20th the Washington Post closed an editorial with these words,

"Here is one more reason to put aside the freeze. Arms controllers can do better." Maybe they could but they have not!

The "reason" the Post had mentioned was the "what if" the US got an agreement in the INF talks in which the Soviets would reduce some of their missiles in Europe while the US could deploy some of its cruise and Pershings. The suggestion, of course, being that this would be a better result than a freeze that stops both sides exactly where they are without the deployment of any new missiles. The Post seems to ignore that such an agreement would be a further proliferation of nuclear weapons, apparently concluding, as we in the US all too often seem to do, that only Soviet increases contribute to the arms race.

If such an agreement were reached there is little doubt that the item the US would deploy is the cruise missile. And if that is true arms control in the future will have been dealt a serious setback because cruise missiles will be the most difficult prime item in the nuclear inventories to verify. Naturally the Soviets will then want some and the spiral of the arms race will be spurred.

The major attribute of the freeze is that it stops the next round of the arms fight, preventing such weapons as the cruise, bigger and more accurate missiles, and new generation nuclear bombers from entering the arsenals of either side. All other arms control approaches currently being discussed fail to achieve that sweeping a limitation on the nuclear arms race.

Arms controllers with their complicated approaches to strategic arms negotiations have had a chance to prove that they can get something better than a freeze for over ten years now but they have failed. SALT 1 and 2 were certainly not better since the arsenals of both sides more than doubled during their negotiation and the agreements allowed build-ups and modern replacements that destabilize the current balance. The START talks clearly haven't even started to stop the arms race, and the INF talks are an obvious non-starter. In fact in all of the arms talks to date all that has been achieved is that each side has given up only that which it decided would not work.

The freeze alone among the options offers a new starting point. It is not meant to be an end-all, but merely a new beginning in which reduction talks could get underway in an improved atmosphere, out from under the gun of constantly increasing threats by each side as their weapons increase or improve.

In spite of the fact that overwhelming arguments to the contrary written by the best strategic planners in the world appear weekly in the Post, the paper seems to be leaning toward support of the president's barrage of misinformation suggesting that the Soviets have acquired some sort of a nuclear edge. Why else argue that a US add-on and a Soviet decrease is better than a mutual freeze. Such an approach represents an illogical understanding of the arms race that is dangerous. The Soviets can not accept an agreement that limits them more than us, especially when in their view they have only just begun to close the gap of US superiority.

It is true there has been a build-up in Soviet nuclear weaponry in Europe, but there is no sound military reason to match that increase, and doing so politically could even prove disastrous. The Soviet SS-20 deployment has taken place because the Soviets were rebuffed in their efforts in both the MBFR and SALT negotiations to put limits or reductions on US and NATO nuclear forces in Europe. That meant that the Soviet Union proper could be attacked by US/Nato forces not limited by arms agreements, while the Soviets were unable to attack the US homeland with such unrestricted forces. Obviously the Soviets decided that was unfair and needed to be dealt with by a force that evened out that Western advantage. Thus the SS-20 deployment.

The US had considered deployment of large numbers of intermediate range missiles in Europe many years ago when the Soviets first began deploying the SS-4 & 5, but we decided, logically I contend, that such a force would be irrelevant. We were sure than any war in Europe would lead to a US/Soviet confrontation and wanted that belief to be perpetuated. In such a case we believed we had more than enough nuclear forces in our more secure US bases, and in the sea based forces near Europe, to destroy the USSR and thus enough to deter such an attack in the first place.

That fact is as true today as then. Our decision was partly based on the fact that the deployments might cause our European allies to think we were separating US defense from European defense, thus undermining the idea that an attack on them would be viewed as an attack on us. It is the apparent changing of that attitude that makes the planned US European deployments so politically explosive. For military perceptions alone it might be wise to match the Soviet theater weaponry, but when the political factors of the anti-nuke movement in Europe, and the perception that the missiles are designed to fight a war limited to Europe are factored in, any military gains, especially such slight gains, are cancelled out. The US, even without the new European deployments, has 20,000 nuclear warheads deliverable on targets in the USSR and Eastern Europe. Additional numbers could only serve to reshuffle the rubble.

But a freeze is a viable idea. Without offering any dangers it could save both sides billions of dollars/rubles and help the US to regain its former status as a peace seeker rather than a militarist nation. The US is incredibly strong and our current nuclear force is deterring war effectively.

It may be true that many congressmen revealed differing understandings of the shape of the freeze during the pre-Easter debate. But that can be mostly explained by the fact that many of them have been newly won over to the freeze by the strength of the support of their constituents. Many of them have hardly had the time to investigate the details, but are still willing to support the freeze now that it is politically viable since they have known for years that the US had a sufficiency in strategic nuclear power, and only supported additional strategic defense expenditures because it was the politically more expedient route.

But the freeze resolution's initiators and leaders have made it quite clear that what they seek is a mutual freeze on all new nuclear deployments to prevent both add ons and replacements. They intend the freeze as a first step toward reductions that can be better achieved when removed from the pressures of ever increasing inventories. If those follow on negotiations drag on far too long, freeze supporters and the frozen parties could easily work out needed replacement provisions but that is certainly something neither anticipated for now, nor needed before the freeze can be endorsed.

In my opinion we need new missiles, at home or in Europe, like we need the return of double digit inflation. And come to think of it, if we get the one we will probably get the other.

4/17/83 Star Wars Mythology

I deliberately waited about reacting to the president's defensive nuclear system call in hopes that later statements would clarify his views and eliminate some, if not all, of my reservations. But that has not happened— the idea seems as irresponsible now as it did the night he first mentioned it.

I do not state those words lightly. For years I was among those troubled by the concept of mutual deterrence and hoped that a defensive system might appear that would make the nuclear nightmare disappear. When the Soviets first began to deploy an ABM system around Vilnius and then Moscow, I was among those who initially resented the US cries that the Soviet move was an escalation of the arms race. Is it not better to build weapons designed to defend than to attack I wondered? And I was pleased when Kosygin used that same tact in one of the first Soviet press conferences with Western reporters. But my support was for a different kind of an ABM, one whose abilities could be limited to a defensive role. Alas, it did not work.

Later when I became privy to far more information about the ABM concept, I became a strong advocate of its elimination. It was proven beyond a doubt to me that it could never give the kind of protection needed to garner security in a nuclear war, and that it would always be enormously easy and far cheaper to overcome even the thickest of the anti-ballistic missile systems with either more war heads or more maneuverability of them. Thus I considered the major achievement of the SALT talks to be that portion of agreement that, for all practical purposes, eliminated ABMs as a factor in the arms race.

But now the president is resuggesting that a defensive approach is the way to get us out of this nuclear era and its threat of the elimination of humankind. Apparently forgetting that when the Soviets made the same claim we saw their acts as destabilizing. Would that he were right but I believe he is not.

First, because I suspect it is all an illusion. Current technology tells us that there is no feasible way to protect ourselves from a nuclear attack other than through a deterrent force that says to any foe, if you attack us you shall perish. Now I am not fool enough to think that current reality is locked into permanence. If we accepted that kind of reasoning there would be no airplanes, no suspension bridges, few cures for what ails many of us, and certainly no effort in space. Science and technology are constantly changing, making possible tomorrow what was impossible today.

But the trends of both science and war tell us that we are more likely to find more awesome ways of killing people in the future than of protecting them. Take, for example, the systems the president seems to be pinning his hopes on for a nuclear defense, lasers, particle beams, and space stations. It is conceivable that some combination of them will be able to sit out in space and destroy incoming missiles—probably not all of them in an all out attack but some portion of them. But more importantly that same system can also most likely launch its death ray at the cities that revolve under it, and if so it becomes an even more frightening offensive weapon than the weapons we designed it allegedly to defeat. And mark my words: if we can build it, eventually so can they. How much safer would we then be? Will not each side then attempt to add more missiles in order to assure they can overwhelm the new defenses?

To me it would be far more intelligent to work toward ways to stop the arms race where it is, than to spur it by research which, although done in the name of defense, will likely lead to ever more deadly weaponry of offense. I suggest the president would be far smarter to accept the freeze, even with the imbalances he argues it has today, than to raise the level of the race he says we trail in already by making it a high priority to acquire even more destabilizing weapons.

I am not suggesting we should not continue to do R&D in the laser, particle beam and other areas. There is not enough trust between the Soviets and ourselves to take such risks. But rather than calling such developments the wave of the future, we would be better served, I believe, if we followed a cautious experimental tract while simultaneously trying to find means of agreement with the other side to rule out such weapons in the future. But before that is possible we need to stop the immediate arms race of those weapons we have both already mastered.

With that goal in mind the president appears to be more of an obstructionist than facilitator. Neither his START, Zero Option, or interim Zero Option proposals have contributed usefully to arms control. All have sought such obvious unilateral advantages for the US that their sincerity is doubted and the suspicion reigns supreme that the president's real goal has been additional arms deployment and a reach for superiority. This last defensive star wars decree comes from the same book. Unfortunately, like his other initiatives, it seems more likely just to perpetuate the arms race than to gain some true advantage for the forces of peace.

4/8/84 # America, The Arms Merchant

Not many years ago Americans considered international arms dealers to be immoral men. Today, consciously under the Reagan leadership, we have become the world's major weapons seller, thus bringing into doubt Reagan's own rhetorical question, "It is not whether God is on our side, but whether we are on his?"

In building nuclear weapons a nation can argue they are simply building a deterrent that will never be used, but the same cannot be argued for conventional weapons. They are indeed the weapons THAT ARE USED. All over this globe by young and old, good and bad, male and female, freedom fighter and terrorist, conventional weapons in ever growing numbers and kill potential are being fired in anger.

In 1982, the last year of available figures, the world spent more dollars on conventional weapons than ever before in its history. And although the great powers naturally spent more than their share, an alarming fact is that the underdeveloped nations increased their conventional weapons acquisition even more rapidly than the great powers. Most of those weapons were supplied to these smaller nations by either the US or the USSR. When one groups the sellers together the US and its allies far out-supplied the USSR and its fellow travelers.

It would be logical to assume that if the arms supply market is booming, there must be some obvious benefits to the seller and buyer. Although the seller clearly expects to not only make money but win friends and influence, the evidence suggests this is often not the case.

The US poured billions of dollars of weapons into Iran in order to assure a friendly government in what we considered a strategic spot, but the result was that Iran turned on us as public enemy number one, the Great Satan, and those weapons are now used all over the Middle East, most often against efforts the US supports.

We gave and sold billions more in weaponry to Israel only to see her use of them devastate Lebanon, a US ally, and drive a wedge between us and our former Arab friends that many argue has made the US a useless giant in that crucial region.

And has even Israel gained from this weapon acquisition? Hardly? Although they now hold greater Israel almost in its entirety, a clearly stated dream of Mr. Begin, he has been driven from office, his nation is more divided than at any time since its inception, and more isolated from the world than at any time since 1948. Additionally inflation caused by the weaponry reliance, and leadership disputes have risen to new heights in this gallant but sadly disoriented nation that once held such promise.

The Soviets found similar results from their weapons sales to Egypt, assistance to China and Albania, and their generosity to Somalia and Afghanistan. The US got the same from its military aid to Ethiopia. We have given more total aid over time to South Korea and Taiwan than any other states, yet neither of them today stand as examples of liberty. Zaire is another example where our guns more support tyrants than provide freedom, and recently we find ourselves supporting new despots in Sudan and Chad with nothing to gain except an alleged defeat of Kaddafy, one we have made a whipping boy to illustrate US decisiveness. In none of these places has stability, liberty, or sincere friendship resulted between seller and buyer. It is more like a junkie-pusher relationship.

Perhaps a productive area to get the US and the USSR talking again on something substantive would be a US initiative to talk about limiting the conventional arms trade. Success could have life saving impact on Central America, Africa, the Middle East and indeed much of the world. Few actions could save more lives or eliminate more delusions about the gains from being a weapons pusher. And since the Soviets have suffered as many failures in this arms bazaar as we, agreement might be possible.

11/14/82 # Arms Race Exaggerations

Official government statements about the US/Soviet military balance fly in the face of the recent vote and poll results supporting the nuclear freeze. How can the public support a freeze if the Soviets are spending double the US percent of GNP on defense expenditures, if their missiles out-number ours by several hundred , if their missiles are bigger and if during the SALT talks we stood pat on missile deployments while they added like mad? In a word the answer is "Misspeak."

Misspeak is a term that has entered our language to soften the meaning of the word "lie." It suggests a more subtle distortion of fact designed to mislead for the good of the cause rather than an outright effort to deceive for personal profit. But our government and in fact most of the governments of this world have been caught in so much misrepresentation that a cumulative effect has occurred and all around the globe the people are choosing to seek out the facts for themselves and blatantly challenging their governments' views. Thus, in Europe the governments support new nuclear deployments while millions of their folk say, "No;" in Japan, Spain, and elsewhere governments are falling and in our capitol city pundits are puzzled by the credibility revolution.

Misspeak is usually the truth but a truth designed to alter a bigger truth that might weaken support for the policy the misspeak was created to protect. For example, yes, the Soviets do spend twice the percentage of their GNP that the US does for defense. But that does not mean, as it is designed to suggest and usually taken to mean, that they spend twice as much money for defense as we do. Since the Soviet GNP is roughly only half as large as ours, it actually means we both spend about the same amount of money on defense. But that more correct statement is not nearly as alarming and helpful in garnering support for increased US military spending, so the more deceitful, even if moderately accurate, phraseology is government preferred.

The even more refined truth is that we really do not know how much the Soviets spend on defense and the figures we freely toss out are only estimates. And even more deceitful is the fact that those estimates, usually accepted by most of the public and the Congress as sacrosanct data, are not estimates of what the Soviets actually spend to field their defense force, but estimates of what we would have to spend at our prices to field their military. To grasp what that means one simply has to realize that since the US gave a raise to our servicemen this year, that raise will appear in next year's calculations of

Soviet military spending, even though they likely will not have raised the salaries of their troops and pay their forces exponentially lower salaries on the average than we do. For example we pay raw recruits about $500 a month and they pay less than $10.

Such misspeak clearly distorts the picture of military expenditures unfairly, but how many of you have been aware of it? Could someone be trying to mislead us?

Another misspeak example can be found in the missile count and myth of US standdown during the SALT process. It is true that the Soviets have a higher aggregate number of strategic missile launchers and bombers than the US, and that when the SALT process began the US had led in that calculation. But during SALT the US deliberately elected to increase its arsenal by the more efficient means of adding warheads per launcher rather than by the more costly addition of new launchers and bombers. The Soviets eventually will do the same, but they developed the technology to add warheads more slowly than we and were thus forced to bring on warheads by the more costly one at a time method. Thus, one could argue they were beating us in the missile race, but in actuality we were increasing our ability to destroy targets faster than they were. Thus, the claims that we were idle while they were running ahead are patently absurd.

And perhaps the worst misspeak of all in this military arena is the scare about giant Soviet missiles accruing to them a greater offensive capability. The US uses this infamous missile size comparison to scare the bejessus out of recalcitrants who develop a worry about excessive US military spending. But in spite of the enormous amounts of dollars allocated to the US Defense Department annually, I hope you have noted that they never use it to build giant missiles the size of the Soviet monsters, except way back in the early '50s when we produced the now aging (and more dangerous to us than the Reds) Atlas and Titans. We built them then only because at that time our technology was at about the same state that Soviet technology has been in the '60s and '70s. We do not do so now because we have moved beyond that rather primitive state of the art and we can build more sophisticated, reliable, dependable, effective, and accurate missiles, almost always smaller, yet better. The Soviet "biggies" threaten us only in our pocketbook by giving the military industrial complex and the politicians a smokescreen for the biggest pork barrel spending of all.

But why does this work so well? Clearly our congressmen either know the reality of the misspeak or have someone on their staffs who can see through it via study. The answer is complex, tied up with our obsessive fear of communism and the US political process. Americans fear the spread of communism and the danger of its expansion via military means so strongly, that political history shows we rather consistently will throw out of office a congressman or a president who appears unwilling to build the weapons systems necessary to keep us stronger than those Red villains. So congressmen perceive that even if they know that a new weapons system really is not needed, their constituents will not know that reality and might fall prey to a challenging politician who will accuse the incumbent of being soft on communism and not supportive of national defense. It's easier to spend the money, and not inconsequentially reap considerable donations from the defense industry, than it is to try and educate their district populaces to the degree necessary to ensure they will not misunderstand the vote and seek a more devoutly anti-communist politician.

Misspeak is something I think we can all understand, the little white lie meant not to do harm but to help. But such lies have led to the weakening of respect for governments all over this globe and in America too. I believe all governments would profit from a turn away from this tactic, from showing more respect for the intellect and integrity of the people they represent. But that will only happen when we, the people, prove to them that we not only can understand reality, but will take the time to garner it with or without their leadership, and then demand that they respond to our informed desires. The rush is on—join it!!

6/27/82

Poor Propagandists

One would think that with all of the practice garnered in the free enterprise advertising system, American leaders would be better at the propaganda game than they seem to be. The recent disarmament conference at the United Nations is a case in point. The leaders of the world's two major powers both had an opportunity to impress the rest of the world with their interest in peace and nuclear disarmament, and it seems clear that in the eyes of the world the USA came in second.

The Russians went before the conference and made two offers, both of which had already captured the interest of the world. They said they were willing to freeze nuclear weapons at current levels while the new US/Soviet talks get underway, and also that they would adopt a "No-First-Use" of nukes policy. This occurred just a few days after the city of New York and the United Nations were rocked by an outstanding demonstration in favor of the freeze by Americans and others from all over the world. The Soviets move obviously caught the tenor of the delegates and gave them something to applaud.

On the other hand the US president went before that august body with a speech reminiscent of the late 1940s. Reagan went to the rostrum carrying the image of a cowboy hip-shooter, and he did nothing to alter that vision. Instead he raged about the Soviet misdeeds since the end of World War II, glossed America's achievements as if we were the perfect nation, belittled the Soviet offers as empty words and then offered nothing in their stead. Well, maybe "nothing" is not fair. He repeated his START proposal and his Zero Option position concerning theater nuclear weapons.

Since Reagan's START proposal would drastically reduce the number of nuclear weapons and the Zero Option the same, why did the delegates only respond lukewarmly to his offers while receiving the Soviet ones with enthusiastic applause? Is it because, as Jean Kirkpatrick suggests, the enemy has seized control of the UN, and that that body is now ruled by representatives of evil men?

I am not very impressed with the UN record, but I do think it provides us with a reasonable gauge of the attitudes of most nations towards the problems of the world. Therefore, if the UN majority opinion is that the Russians are more interested in realistic arms control than the United States, something is wrong somewhere.

Let's examine the Reagan and Brezhnev proposals and see where the fault might lie. The Russian freeze offer would result in an immediate halt to the deployment and testing of any new nuclear weapons not already in place, such limits to last through the duration of the planned START talks. On the other hand the Reagan proposal will allow both sides to deploy virtually anything and everything they have underway or are considering for the duration of those talks. For the US that means we can put out 2,000 more warheads via the MX, add the Stealth and B-1 bombers to our forces and continue to replace our older SLBM's with Trident missiles. It is not likely that the Soviets would sit idly by while we did so, thus resulting in a situation like the earlier SALT talks in which each side more than doubled its forces while talking about arms control. Everyone I know believes that the START talks will be the most difficult to negotiate so far, so the talks are sure to be long. Many cynics—there are lots of those at the United Nations—believe that is exactly what Reagan has in mind. Beneath the surface, then, it is easy to see why the Soviet proposal looks more helpful toward containing the arms race.

The Soviet "No-First-Use" offer is less significant than their freeze suggestion because, as was quickly shown by the US response, who believes that such a promise made in times of peace will necessarily be adhered to in times of stress? Still it was a shrewd propaganda move that makes the Soviets look more inclined to avoid a nuclear war than the United States. We should have responded in kind, for there is nothing to lose in so doing. The Soviets will not believe us any more than we do them, but by both agreeing we look better to the world.

Reagan's Zero Option is a non-starter with the Soviets and viewed by the world as simplistic. Anyone at all informed about nuclear matters, and most UN delegates are, knows that what we are suggesting when stripped of all the zero glitter, is that the Russians give up what they already have and just spent billions to deploy, while we would only forego what we were planning to do, assuming Congress lets us and our allies accept them. Both of those "maybes" are far from certain. Additionally, such informed people also know that the Soviet intermediate force which Reagan wishes to dismantle through his Zero Option was designed to counter a long term US/NATO advantage in Central Europe in nuclear weaponry—an advantage that the US had refused to let the Soviets cope with during the SALT negotiations, although they had tried to do so since SALT's inception. Finally the Soviets gave up at the table and did something about that Western advantage by the SS-20 deployments. Now, via Zero Option, we wish to return to the status-quo-ante rather than admit it was in our edge.

And finally we have Haig's shabby response to the Soviet propaganda victory. Last Sunday he tried to belittle the Soviet offers by saying they were insincere. For proof he revealed a series of Soviet missile test firings, stating that they were at an unprecedented level. Of course the Soviets never suggested they might unilaterally freeze, and there has certainly been no decrease in US testing activity because of our START proposal. In fact the new administration has promised just the opposite. It seems to me we might be impressed by the increase in Soviet tests after such a freeze offer, but for the opposite reason. It might be an indication that they told their military to test as much as they can and as fast as they can, because if the US takes up their offer, there will be a freeze!

Come on leaders, let's get with it and stop letting the bad guys confuse the world about who wears the black hats. Folks should not need a scorecard.

CHAPTER THREE

Serendipity

The subject of international affairs is so serious and at times frustrating that one simply has to have a sense of humor to endure it. Thus, I frequently tried to insert a bit of humor in my columns to keep the readers from deciding it was too serious and stern to hold their attention. Buchwald or Baker I am not, but many of these attempts at levity drew a substantial reader response.

The one about the weekly prayer meeting of the Joint Chiefs and friends merited a phone call before I was even out of bed the Sunday it appeared. The reader had been delighted and made my day by telling me I had made hers.

The one on dog poop, yes you read it correctly, "dog poop," set me up for an obvious rejoinder. I was warned but chirped back, "they wouldn't dare." But of course they did, "we suspected all along Mr. Clark was full of such and at last he confirms it for us." That was the general theme of numerous letters. One reader happened to see me on campus the next day. We had never met, and to this day I do not know his name, but as we approached a door from opposite directions and nearly simultaneously passed through, he smiled and said, "Ruff, Ruff." After you read it you will understand why I broke up as we parted. Membership is still available, in fact the club is yet quite exclusive.

The Academy Award bit is my favorite. I have yet to make it through one of those shows without falling asleep. But since it is true confession time, I have to admit, I really did not have the nightmare so recorded.

I also like the one about the Soviet sins in Central America. THE WASHINGTON POST WEEKLY likes to run its government memo of the week and once featured one in which President Reagan classified the nation of Mali as one to whom American military aid could be given since it would enhance US security. Obviously, the concept of national security has gotten a little out of hand when Mali can save it and Nicaragua threaten it. By the way the USA is doing something about the noted scholarship gap as we are now offering more to the young folk of Central America. It may be a bit late, but it sure proves that some good can come out of almost anything, including Secretary Michel's rather unintended lampoonish speech.

I frequently tell the students in my classes that every action on the world scene produces both plusses and minuses, helps some, hurts others and does some harm and some good depending on viewpoints and positions. I should add that all of it has its serious side and its levity, if one is just able to appreciate the humor. Sometimes however, the humor is so black it is hard to laugh at it without simultaneously getting stomach cramps.

Have at it, and if you come up with any gems of satire on current events pass them on to me. I love 'em!

Foreign Fun

People all over the world enjoy humor and it seems especially important to them when times are tough. Humor is cultural and tells us a lot about the conditions that prevail where the joke, cartoon, or essay originates. I thought it might be useful for you readers if I retold some of the best of the current crop of humor from yonland.

The first story most recently came out of China but it has appeared in similar form all over the globe wherever supreme power has rested, usually precariously, in the hands of a few. It goes: Three prisoners are bemoaning their fates to one another, hungry, tired and shivering in their tattered clothes on a cold wintry morning. The first says, " I should have known better, I became a supporter of Comrade X." The second proffers an ironic smirk, "What a crazy world—I was an opponent of Comrade X." The third then waves his hand helplessly, "You two should feel bad, I am Comrade X."

Hungarian: A slightly similar story is popular in Budapest. Two high ranking party officials make a tour of the public institutions and wherever they go, the ranking official demands that the institutional director reduce the costs of his operation by a 1/4, 1/2, or even more, even at schools and hospitals. Their last stop of the day is at a prison and there, much to the surprise of #2, #1 insists that the warden raise his costs per prisoner 400 fold. As they leave the subordinate's puzzlement overcomes him and he asks why they should decrease funding for pupils and patients but increase it for cursed prisoners? The leader confidently replies: "You don't think we'll be sent to a school or hospital when they kick us out do you?"

Russian: The drinking problem in the Soviet Union has become notorious and the Kremlin's leaders have tried everything short of prohibition, which they fear might lead to a revolution, to stop or at least hide the fact. When Nixon last visited the USSR, he undiplomatically asked Brezhnev, "Where are all the drunks?" Brezhnev responded that the rumors of rampant alcoholism were unfounded, indeed only propaganda perpetuated by the capitalists. "If all the world were communists we would have no drunks," he boasted and to prove it offered to drive through the streets of Moscow. "If you can find a drunk, you can shoot him," he said as he pushed a weapon into Nixon's hands. Naturally, Brezhnev then took the precaution of having the KGB drive ahead of him and clear the streets of any drunks. Nixon was impressed with the appearance of sobriety in the city.

Later when Brezhnev visited Nixon in DC, after a few drinks he challenged Nixon to match his offer and the reluctant president felt impelled to agree. They drove through the streets and after a few blocks spotted six loudly singing, carousing and boisterous men, clearly on a drunken spree. Before Nixon could prevent it, Brezhnev leaped from the car, shot them all and the car sped away. The next morning the WASHINGTON POST headline read "UNKNOWN MADMAN KILLS SIX MEMBERS OF SOVIET EMBASSY STAFF!"

England: The great humor magazine, *PUNCH*, during the middle of the

US political campaign featured a cartoon showing three faces behind glass jars. Carter stood behind a jar marked "peanuts," Reagan's jar was labeled "gun drops," and the third face, wearing a hat identifying it as the US electorate, was fronted by a jar of all-day suckers.

TURKEY: This is a true story about embassy life. A good friend of mine was hosting a dinner party in Moscow, USSR, and greeting each arriving couple with a handshake for the males and a kiss on the cheek for the females. When a Turkish colonel and his lovely wife arrived, the colonel's face visibly reddened at the affection bestowed upon his wife by the host. The American noticed, flashed an embarrassed smile, and remarked, "an old American custom." The Turk, still a bit taken aback, replied, "Yes, now an old Turkish custom, I must kill you." The later arriving wives missed out on the kisses.

Mexico: The humor of Mexico these days seldom portrays "Gringoland" in a very good light and usually concentrates on America's new found and rather clumsy interest in Mexico as a result of their ample supply of oil. One Mexico City cartoon clearly expressed a pox on all of our leaders by illustrating a mountain with several ever declining level spots and Uncle Sam tumbling down past each from the high spot marked "Johnson" to lower ones labeled "Nixon," and "Carter" and then off into a void marked "Reagan."

Germany: One of my recent favorites from the German press appeared right after the Soviet invasion of Afghanistan and was a sign in a tourist agency saying, "Visit the USSR—Before It Visits You!"

Russia: Russia's best selling humor magazine, *Krokodil* (Crocodile), and its cartoons tell us a lot about Soviet society. An oldie but goodie came out in 1967 when the Soviets were preparing their fifty years of Communism celebration. Among the new achievements they were to celebrate was a new edict at last placing much of the USSR's work force on a five day week, instead of six (note this was 1967). A slightly inebriated Soviet worker, having just learned the good news, arrives home on a Friday afternoon, late as usual, flings open the door, faces his wife and while waving a bottle of vodka in each hand over his head shouts, "Look dear, two days off!"

A more recent *Krokodil* cartoon showed two naked Russian youth, obviously representing Adam and Eve, who were rushing toward an apple tree, but the tree held no fruit, merely blue jeans, tape recorders and rock record albums. Russian youth seem to have grown very tired of the communist polemics about a new life of shared mediocrity and are demanding and seeking the material goods of the "decadent West."

Speaking of foreign countries my favorite Texas cartoon shows the wall of the Texas State Fair with an absolutely huge watermelon laying next to it. Out of the watermelon sticks a sign and the words "second prize."

I'll close with this bit of insight from my youngest son. He was five years old when we returned from our tour at the US Embassy in the USSR, and one night as I was getting ready to go out and give another of the many talks I gave to civic groups about that mysterious land, he was sharing the bathroom with me. I was shaving while he was sitting on the throne. "Where are you going tonight, Dad?" he asked pensively.

"To give another talk on Russia," I replied, and then with what I thought was a rare moment of genius I added, "Say, you lived in the USSR for two years. What do you think I should tell them about it?"

His eyes got big and it was clear he was analyzing the last two years via a fast rewind of his mind. Then he spoke with great confidence, "I think I would talk about Copenhagen." We had visited a mere three days in that lovely and quaint Danish city compared to two full years in the USSR, but Darren was right. Any one in his right mind would rather hear about that bright, cheery, and happy old capitalist city than the drabness of the world's first and foremost communnist state. From the mouths of babes and humorists.

<table><tr><td>3/13/83</td><td align="right"># Hither and Yon</td></tr></table>

"They are not really all that informed about world affairs," the teacher said. "In fact on a global information test they said things like El Salvador and Costa Rica are in Africa, that malaria is caused by snails that harbor in the wet rice paddies, and that Khomeni is a communist dictator."

Now groups like that make me feel pretty confident so I opened my talk with a sense of optimism and a feeling of assurance that I would enlighten the class's knowledge and interest in world affairs. I should have known better.

Taking advantage of the reported misinformation about malaria I explained that it is a most debilitating disease in much of the world, but that most are not aware that humans are the second choice of the malaria carrying mosquitoes who transmit it by their bite. They actually prefer to bite the ubiquitous, in most of the world, water buffalo, but unfortunately for humankind during the monsoon season in most of the developing world, the water buffaloes love to spend their non-working time wallowing in deep pools of rain water and mud. The mud gets all over their bodies and the mosquitoes are turned off and look for secondary targets. If we could just teach people to hose off their water buffaloes we could reduce the incidence of malaria significantly, and without harm to the buffaloes since they handle the mosquito bite with little ill effect.

But instead of that interesting anecdote impressing my young charges, it only illicited some weird glances and several water buffalo jokes.

So I switched the subject to Costa Rica and hoped to impress them with that country's ingenuity by noting that a Costa Rican company had discovered it could meet a cattle feed need, at a much lower price, through coffee, without cream and sugar of course. They had found that they could feed the cattle the leftover ruffage from the coffee powder process and with no side effects since the caffeine and the tannin went with the main product, not the ruffage.

"Yeah, well if you know so much about the world explain to us what is going on in Angola," responded a very precocious young girl. It was the beginning of the end, but I had to accept the challenge.

Angola is a communist run country near South Africa and next to a territory called Namibia which is seeking its freedom from South Africa.

"Isn't that where the Gestapo is working?"

" NO, I think you mean SWAPO. SWAPO is the name of the most effective of the guerrilla groups seeking independence for Namibia. The South Africans accuse Angola of helping SWAPO and as a result they often invade Angola's southern areas and make military sweeps or air raids."

"I think you have South Africa and Israel mixed up. It's Israel that has invaded Lebanon's southern region . . . ."

"No," I interrupted, "I am not mixed up although the cases are a bit similar. You see Angola has a lot of different fighting groups that oppose its current government. One of them has set up its headquarters in the south and is protected by the South Africans."

"Are you sure you are not mixed up—isn't it Major Haddad you are talking about?"

My tone grew a little sharp, weren't these kids supposed to be uninformed? But I held on and explained that Haddad was indeed also a factional commander, but that he operated in southern Lebanon and was supported by Israel, while I was referring to a man named Savambi, who was supported by the South Africans in southern Angola. I went on to note that he had formerly been supported by the USA in his bid to rule Angola.

"But wasn't he a communist?"

I explained that although earlier he had called himself a Red, he was somehow less red now than those the Soviets had helped to take charge in Angola with the help of the Cubans. Mentioning the Cubans was a big mistake.

Naturally, they wanted to know what the Cubans were doing there. My answer that mostly they were guarding US owned oil fields to protect them from attacks by communist guerrilla forces just got me in deeper and the whole issue became confused even more.

Quite a bit later I think it all got sorted out—mosquitoes prefer coffee to humans, and Angola and Lebanon have challenged Israel and South Africa to a water buffalo race in Cuba, or something like that. All clear?

2/1/81

This Crazy World

If all the world's a stage than clearly the current booking is a tragi-comedy, a theater of the absurd. One merely has to observe world-wide events to see the humor a la Buchwald in them. Yet, the laughter often turns to tears when one realizes this joke has tragic results that can cause millions to starve, hundreds of thousands to be killed and tortured, and so many many more to be locked into despair—a despair for freedom or economic security. Some examples follow.

The term SALT stands for Strategic Arms Reduction Talks, so clearly their goal was to limit or reduce the number of nuclear weapons. A SALT One Agreement was reached after more than three years of tedious negotiations during which strangely the two parties concerned *increased* their nuclear warhead numbers by some 50 percent. But the real ironic twist is that when the treaty came before the US Senate, the administration had to promise the senators it would INCREASE MILITARY SPENDING to compensate for the treaty before they felt free to ratify it. SALT Two took eight more years to negotiate and in that time the nuclear warhead total of the two parties more than doubled, but when that treaty came up for ratification the same war dance began all over again. In order to ratify, the senators insisted the US would have to promise to build the MX or the cruise missile or both. Thus, in today's world it is obvious arms control's major purpose is to ensure arms growth. Absurd?

Most of the world's leaders like to speak of their peace accomplishments, yet since the "Great War," (you know, ol' WW II) there has been a succession of smaller wars which have together far exceeded the alleged "biggie" in magnitude of firepower, years of engagement, and weaponry enhancement. North Korea against South Korea, North Vietnam fighting South Vietnam, North Yemen shooting at South Yemen, and China, in the north, shooting at Vietnam to its south. Is there something in the water that makes northerners hate southerners and vice versa? Why can't all of those backward, uneducated people learn from the USA example, whoever heard of Americans fighting between the north and south?

The world must appear terribly out of kilter and absurd to communist observers also. One of their basic tenets is that capitalism is the cause of wars. They believe that if all countries were ruled, as is their myth, by the working class, there would be no exploitation and therefore no wars. Yet almost all the most recent wars are between so called communists. Vietnam has invaded Cambodia, perhaps the closest society ever to the communist manifesto as ruled by Pol Pot, and China has attacked Vietnam in retaliation for its attack on a communist brother. In Africa Somalia and Ethiopia, both claiming to be Socialist People's Republics are flinging modern spears from rocket launchers at one another and catching the USA and the USSR in a near comic-opera double take that made them switch sides in that battle almost in mid-stream.

And perhaps the most absurd case of all is in Angola where three communist espousing rebel groups, respectively supported by the Soviets, US, and the Chinese are so embroiled in a bitter struggle that it provides this bitter twist. Cuban troops entered on the Soviet side are now primarily used to provide protection for US oil company owned facilities against attacks by the other once admitted communist groups who are supplied by the USA and China. And we used to wonder why US presidents turned grey while in office.

The home scene does not escape this comedic performance either. In 1974 a Republican president was hounded out of office in disgrace for his misuse of power. His abuse of human rights, via the CIA excesses, the Watergate affair, and the prolongation of an unpopular war, were alleged to have taught

America a lesson on the limits of power, and many predicted that the Republicans had suffered such a severe blow that they might never recover.

Yet only six years later that party has returned to power in a near landslide and the winners are curiously filling their ranks with men who cut their teeth in that Nixonian era. Men who are now telling us we must put the lessons of those days behind us and return to the pragmatic balance of power concepts that caused those miseries. Is pragmatism better? The experienced, pragmatic team of Nixon, Kissinger and Haig publicly showed an interest in the following clearly identified goals: A SALT Two Agreement with the Soviets, no US soldiers in combat, an Israeli/Egyptian peace settlement, full and normal relations with the PRC, a signed and ratified Panama Canal Treaty, and a clear link between detente and Soviet actions towards others. But in spite of having a Ph.D in foreign affairs at the helm, and pursuing the pragmatic approach they called "hard ball," they failed totally on all of the above.

Then along came an inexperienced peanut farmer with a pea brained brother, a toothy smile, and a clearly naive emphasis on human rights over balance of power and lo and behold, virtually unnoticed and clearly without any panache, he accomplished all six of the above listed objectives. His reward? Total rejection by the US electorate and the other leading statesmen of the world.

Stay tuned folks: the only thing we can be sure of is that it will probably get even wilder. Why, who knows, some day we may even have a middle eastern mystic potentate hold a superpower at bay by seizing hostages from the sanctity of their embassy and demanding the return of an exiled ruler who is hiding out in a third country. Or even more absurd we might even hear from a conservative presidential candidate, who promises to balance the budget by cutting welfare spending and increasing the military budget even more. We might believe him, elect him, encourage him to do it, end up with the greatest deficit in the world's history and reelect him again just to prove we meant it. Oh come on Don—that's too absurd—OR IS IT?

1/13/85

Times Are Ruff

I've been doing a lot of walking lately, and since this is the time of the year when my home is a skier's paradise, I spend a lot of time looking down in order to avoid those slippery icy sections that seem to put more folks in a cast than the slopes do. But much to my surprise I'm dodging more dog puckey of multitudinous sizes, shapes, and colors than I am icy spots. It must be that one does not notice the droppings of our pet friends as much in the snow-free times of the year. That yucky stuff hides in the grass and contrasts less when there is no snow cover. Now I've been a dog owner and lover for years, although admittedly when my last one died after some 16 years I grieved less than in the past. And I know all about that American dream of a child and his/her dog. They look great in the story books and on the movie screen, but the reality is quite different—those hairy critters are downright dirty and unsanitary. The combination of dog and kid probably spreads more disease than any other matched pair in our otherwise sanitation wise land.

I say it is time we did something about it. All of you get out, take a walk, and if you see what I've seen let's join together and form a new political movement called RUFF. RUFF stands for Reduce Unnecessary Fertilizing Feces.

RUFF will not be unfair. I know some folks simply can't get along without their dogs. Hunters need dogs to add to the already poop-filled woods, the blind need their seeing eye friends, and the military and cops need dogs to sniff out the enemy and locate hidden drug shipments—all worthy achievements. So we have to find a way to let dogs remain in our society. But that does not mean we cannot shape them up. After all we are always being told how noble and smart they are, so why not potty train the characters. We punish parents and ostracize kids who can't learn where and when to poop, so why not dogs as well?

Of course we have to give puppies a breathing period. They are so darn cute, you can't help but love them even if they do chew your shoes, soil your favorite rug, carpet, or bedspread, and bark or whine all night making your neighbors hate you, until they see the furry little sweethearts. So we need to have reasonable rules.

I suggest that the first year dogs be on their own, free to roam and leave their little territorial claims wherever they wish. But after one year, they should be required to go for a license, and if they cannot pass a test, you know like a driver's exam, where they demonstrate that they poop only in toilets, they should get stuck with a conditional license only. By the next year, their second, which is the equivalent of a fourteen year old human, they have to pass the test or be put away in cement and stainless steel jails until they can learn civilized habits and use the john. Recalcitrants who never learn will be doomed to life imprisonment for the safety, sanity, and sanitation of the rest of us. By the way, just to prove we are not discriminating all fourteen year old humans should have to pass the same test. I'm a bit suspicious about some of those piles I've seen.

If the dogs can't hack it, there are, of course, other alternatives. Street washers with jet streams could flood the town everyday. We will have to eliminate grass, but for a lot of us mowers that would not be a total loss. Or really nice dogs who have failed to pass the exam could be required to wear diapers, that would do a lot for the economy by spurring the disposable diaper companies to bigger and better things. I can just picture some of those darling TV commercials now.

I'm sure there are a lot of other solutions as well, so if you think of any "Jim Dandies" just send them on to me in the mail along with your RUFF MEMBERSHIP APPLICATION.

Now if all this sounds a bit far out to you, just compare it to the Reagan Star Wars solution for ridding the world of the fear of nuclear war, add to it the spectre of Reagan now telling us he really wants an arms control agreement with the evil empire, and compare that to his military budget, rejection of a nuclear freeze, and support of the Contras in Nicaragua and it will all make sense. "Ruff, Ruff!!"

God's On Our Side?

It was one of those every other Fridays and thus time for a meeting of the "Old Testament is Supreme Club" in Washington, DC. The Reverend Harry Standill and his favorite congressmen, Senators Nehemiah Sherman, and Hesse Sterns were in a joyful mood for clearly their brand of Christianity was sweeping the nation's capitol. The Joint Chiefs were holding one of their bimonthly meetings for prayer, Bible study, and Christian fellowship and all over the globe the nation was clearly doing, in their minds, whatever God wanted us to do. A window was open and I overheard their conversation.

They spoke of the glorious display of God's might via gunboat diplomacy in Central America where things were finally beginning to turn the tide against Godless communism. They praised the USA sponsored military intrusion of Nicarauga where the sinful Sandanista were now fearfully backtracking. And spoke in glee about Guatamala where Rios Montt was working his evangelical miracles on the Indian tribes and bringing peace out of the gun barrel.

They gloated that the US Congress had awakened from its sleep and was rearming at an acceptable, if slower than preferred, rate, and that the MX, oh glory, glory halleluiah, would be built. Yes and things looked very good even for torpedoeing the arms talks with the, Lord forgive me for mentioning the name without an epitaph, COMMUNISTS.

Americans were also once again giving proper support to the gallant Israelis who were holding the fort for Yahweh and his plan of democracy in the Middle East. They had secured once again the rightful domination of a Christian minority over most of Lebanon, or at least those parts that were not preordained to return to the rightful owners, the Jews, along with the West Bank and Gaza. And the Israeli, through the enlightened leadership of Begin, were once again proving the strength and power of our God.

Even America was getting gallantly into the act on this battlefield of religions. US troops were helping to secure the Christian control of Lebanon, the right kinds of Christians of course, not the Druze and other heretics, and US soldiers, God Bless 'em, had just entered Chad to teach the inept unbelievers there how to stop Kaddafy with Red Eye missiles and US training. Of course it may take awhile to sort out who should rule in Chad, things have been rather chaotic there for several centuries, but it would be made clearer daily just as soon as we learned which side had gotten help from the commies and which had not. That approach had certainly worked well in the Congo and now the people there lived in Christian, free enterprise bliss.

"Things are going well," Harry smiled and said to Nehemiah. "But not well enough," came the stern reply. "The madmen in the House just passed a resolution cutting off aid to the former Somoza contras now gallantly fighting to bring their nation back into the Christian fold. Can't they understand that those people have been redeemed?"

"And 10 percent of our youth are still not registering for the draft while even more are using birth control pills and not telling their parents," chimed in Senator Sterns. "Congress has rejected several of our nominees for important commissions just because of their Christian resistance to racial mixing, and there is much work to be done. We have not yet begun to roll the commies back in Eastern Europe, and need more guns and troops if we are to drive them into bankruptcy trying to keep up with our military expenditures."

"But that is not the worst of it," solemnly added Sherman, "there is a growing number opposing our policies, and suggesting that we misuse the Good Book in our support. They are saying we should be emphasizing the New Testament and Jesus's teachings."

"Boy are they dumb, his teachings won't work in the real world—they are only for use after we clean out the Reds and their sympathizers and dupes. Let me tell you the plan."

And then someone closed the window and I heard no more, but I'm not worried for they are Christians and only do what is right and just. Right???

11/4/84   # Not Voting Is The Thing To Do

Tired of hearing a thousand and one reasons why you should vote on Tuesday? Me too, so let's get even and not vote.

Why? Because voting is communist. In the average communist country almost everyone votes, well over 95% of the populace does so regularly, although since they only have one choice it's not as tough on their minds as it is for us—don't you just hate having to make decisions?

Non-voting, on the other hand, is obviously apple pie and Chevrolet. No US president has ever received even forty percent of the possible votes since a really good turnout for us on a national election is just over fifty per cent and far less in local elections. So prove you are not one of those Red pawns, a commie lover or sympathizer and stay home Tuesday—those Wheel of Fortune reruns are far more fun and less taxing than voting lines.

And if that's not reason enough to malinger on Tuesday then try this. You can't vote for Reagan or Mondale. Haven't you read the small print? You actually only elect some jerk none of us have ever heard of before or will again. You choose some male or female elector, who belongs to an unaccredited college, THE ELECTORAL COLLEGE, a school that doesn't even have a football team, who then gets to go to Helena and vote for whomever he/she darn well pleases. You don't think so— well it happens almost every election. Do you know that three times those geeks have elected a man that got fewer votes than his opponent? Voting just allows those temporary bureaucrats to run our lives.

Furthermore, on several occasions those electors couldn't do their job right so, hold on to your hats, they let Congress choose the president—remember Congress, the same guys who gave us stumbling Jerry Ford for a Chief Executive.

Still don't believe me, eh? Too brain washed by all those appeals from Howard K. Smith and Paul Newman, do your duty and vote, vote, VOTE.

Hogwash!!! What about this angle? Want to reduce government spending and that obscene deficit? You can do more than Reagan has for a balanced budget by just not voting? Think how much money we could save as a nation if everybody stayed home and didn't vote! No ballots to count, no arguments and law suits over lost or stuffed ballot boxes, dead voters, fake registrations, no poll watchers wasting time when they could be doing something productive, like flossing their teeth.

If the idea caught on we could even eliminate those expensive and boring campaigns and just let the pollsters name the government. If no one shows up at the polls, they will have to assume Ranger Ron has won again, and instead of having to flip channels Tuesday night and staying up late to confirm what we already know, it will all be over and we can see what we want to see, the A Team and wrestling.

And finally don't vote cuz no matter what you do you will be sorry. Any intelligent soul would have to decide to vote for Mondale on just the issue of nuclear weapons alone. He says he will stop building them since we already have ten times more than we need. But RR says we have to build a few thousand more in order to stop, and in the meantime he wants to spend billions more on a new space defense that later he will turn over to the "Evil Empire." If you believe that, leave me your number, for I have a bridge to sell.

So if you vote for Mondale maybe you will help avoid Armegeddon, but don't think you will be appreciated. All of those fat cat employees of the military industrial complex will swear you cost them their jobs, and we won't be able to properly supply Marcos, Pinochet, and Mobutu with the guns they need to preserve their democratic paradises.

While if you do the irrational and vote for the man of steel, just as the economy begins to boom, the boom might get louder and turn into a mushroom cloud. Well, in that case, at least it's not likely anyone will be around to blame you.

So not voting is the only logical, American, anti-communist thing to do. It's tough duty but somebody has to answer to the higher call. Stay home, or even at the office, but above all DON'T VOTE!!!

(Editor's Note: Don Clark was last seen babbling election day slogans as he was shoved into the loony wagon and solemnly pronounced a victim of electionitis—he hopes the absurdity of the above will ensure you all vote your conscience, whatever it is, on election day.)

Dream Awards

Like many of you, I suspect, I tried to watch the Academy Awards Show the other night but fell asleep somewhere between Jack Nicholson and Best Picture. I don't know about you but I had a dream that went something like this.

The auditorium looked just like the one in Hollywood but the audience was composed of members not of the Academy of Motion Pictures but of World Government Officials. There was President Mitterrand, tuxedoed and ready to present an award. He spoke, "The nominees for Best International Hypocrisy are: The Ayatollah Khomeini, Persian Productions and star of DEFEAT OF THE GREAT SATAN; Saddam Hussein, Iraqi Inc., in STEP ON THE GAS; Konstantin Chernenko, Red Star Limited, for THE SECOND TIME AROUND; and Ronald Reagan, Stars and Stripes Video, in SLAP LEATHER YOU COMMIES. The envelope please."

Then a commentator, apparently overlooking the whole affair, came on screen and began showing excerpts of the various nominees, activities, war scenes, speeches, etc. He commented, "Ordinarily we would have expected the Soviet leader to be the real favorite in this category since he represents a company that is renowned for its savagery, but he is such a newcomer on the scene and has been upstaged so much both at home and abroad, we can only consider him a dark horse this year. Undoubtedly his company will win numerous awards such as THE LARGEST NUMBER OF POLITICAL PRISONERS, MOST EFFECTIVE SILENCERS OF POLITICAL DIS-SENT, and other less prestigious and technical categories like EAVESDROPPING, but in the top award area, I just do not expect him to be the winner.

The Ayatollah is, of course, another story. He has, according to him, Allah on his side, and he is even older than Chernenko, old age has often been an advantage; also he was nominated previously for that major production, BLINDFOLD THE AMERICANS, of a few years back. But on the negative side a lot of folks figure Allah won't let him die so there is no hurry, and he has not been able to outshine Iraqi Inc. or even Syria Syndicated in that Middle East market. The Academy likes winners. Additionally he has angered a lot of the academy members from the West, the largest group allowed to vote, by confusing them with his politics. The voters are simply not used to rating hypocrites highly if they are not communist, so until he finally reveals his true red color, the conservative vote won't support him.

Saadam is not really a serious threat either because the voters are usually turned off by losers and he is perceived as such. Here is a guy, who, although a great actor, handsome and almost Rudolph Valentinoish, had a real chance at success after his surprize attack on a distracted and apparently vulnerable Iran, but he offered very little in the way of ridiculous cover story, and then even bogged down. Later he was driven back into his own territory and has not been able to defeat the Iranians, even though they are reportedly reduced to sending kids against him. Even his alleged gas attack has been negated by the lack of any advantage gained, and besides folks remember how humiliated and helpless he was when the Israeli bombed his favorite toy, the nuclear reactor. Underdogs can't win this award. So this brings us to my favorite to win, the only western nominee, although I assure you the academy overlooked some other first rate contenders, Ronnie Reagan. He too has the advantage of his dotage years, although at first he was a real longshot. All he could boast was that he promised to end inflation without increasing unemployment, to reduce government spending and balance the budget, but failed. And his raising unemployment to a new high since the depression was almost canceled out because he delivered on the inflation reduction. The deficit deal kept him in the running though, attaining the highest ever when he had vowed for years it was the major sin of government. Then his arms control gimmickry nosed him out of the pack. Who else had ever disdained a popular arms freeze by insisting that only deep reductions was an acceptable goal, then to argue that such cuts could only be attained by a slight upward adjustment, (about a 1300 warheads), and who else ever so blatantly proved he was not serious by appointing known anti-arms controllers as his chief arms negotiators?

Still this could hardly top the Soviet continued ruthless occupation of Afghanistan so Ranger Ron remained a long shot while the KGB Kid, Andropov, was around. But then Stars and Stripes Productions rang up a box office winner with its BONZO INVADES GRENADA. Now that was hypocrisy to admire. Reagan even asserted that the mighty US invasion of that tiny tourist isle, whose sin was the construction of a tourist airport, was proof that America was back, standing tall, and once again playing a major role in world affairs.

But the clincher that made RR the favorite, in my mind, was the war against Nicaragua and the mining of ports there. As hypocrisy goes this was a winner, especially when accompanied by a simultaneous campaign to return his nation to the Christian fold by allowing prayer in school. Reagan asserted the country had been going to hell ever since prayer in school was outlawed and darned if he didn't seem out to prove it by his Central American violations of laws and basic decency. Wait a minute, I think he is about to announce the winner, back to the show."

"The winner is....!!!!" Heck. I woke up and it was TERMS OF ENDEARMENT.

St. Nick And The MX

12/25/82

> Tis the day after Christmas and all through my house,
> There's plenty of cheering about the roar of the mouse.
> For in spite of the big guns the president could muster,
> Congress has wounded his MX bluster.
> "Ouch," cried the president—"You guys must be asleep. For

without an MX my START is doomed to defeat."

Senator Hollings replies, "Not so, Ho Ho, for without the MX, we may be able to get the economy to go."

"But the Reds are too strong,"the president replies, "We have to rearm to gain respect in their eyes.

"We need not only the MX but the B-1 too, the Pershing, the Trident and the Cruise—Just a bit more and then we can't lose.

"The Russians will learn they can't push us around. No more Polands, Irans or Nicaraugas will abound.

"The Russian missiles are bigger than ours, they are more powerful and too threatening to our stars, Like the Minuteman that just sits under the ground, waiting to be hit and blown all around.

"We need to close the window of vulnerability, to balance the SS-20 and acquire the zero option stability; to deploy the Peacemaker and again become strong.

"But we cannot do it if the "Pinkos" and the naive are influenced by the Reds and won't go along."

But what Ho did happen in that mighty room where the House and the Senate gathered to fume?

The voices of wise men like Kennan, Maxwell Taylor, Hans Bethe, et. al., carried the day and the MX did fall.

"The US weak?" They chuckled and laughed. "Why not even the Russians believe that misspeak.

"Our weapons are more numerous, more accurate and reliable. The MX gambit is just not viable.

10,000, 20,000, 30,000 and more, we already have enough to prevent a war.

"Only madmen would challenge the American bird, not even the Russians can be that absurd!"

Thus a lame duck turns on its Boss; it gives him his gas tax but a much greater loss. His false dream of a return to the arrogance of might has been brought short by reality and greater insight.

And the reasons are so very very good, for Congress would only act so defiantly if they knew that you and I wished that they would.

The work of so many to inform the rest, that more nuclear weapons is the wrong litmus test, has succeeded beyond the dream, and created the opportunity for a peaceful scheme.

It is the realists, not the Hawks or the Doves, who carried this day. Thanks to them, now on this earth our children might stay.

Thank you St. Nick, your present has brought this world good cheer. For without the MX we can reduce the fear.

So I wish you all a Happy New Year, and hope that it brings

with it a nuclear freeze,

That will not only save us money so the hungry can again afford cheeze—but bring with it to this world, at last, a hopeful and peaceful breeze.

Those Insidious Soviets

It probably comes as no surprize to most readers to learn that the Reagan administration is a trifle upset by Soviet fishing in the troubled waters of our backyard, Central America. Our present leadership has fought and lost a recent battle with Congress over aid to the "Contras," whom we claim are freedom fighters trying to restore liberty to Nicaragua where a Soviet-Cuban inspired regime seized power while Jimmy Carter was asleep at the helm of hemispheric security. Thus, in anger, spite, or justifiable prudence the president decided to deal with the Sandinistas his own way, via an economic embargo.

In order to impose such restrictions the president had to indicate that Nicaragua was a threat to our national security. Since Nicaragua is a nation of under 3 million, incredibly poor, barely able to stand up to an attack by 15,000 US paid mercenaries—ooops, freedom fighters; the credibility of that threat is about as sound as the excuse we used earlier to avoid being subject to their charges against us in the World Court for mining their only viable harbor.

So being rather dimwitted I decided to probe further and see just what it is about the Sandinistas that threatens us. I did not have to look far for not only are these people friendly with and receiving economic and military aid from the USSR, the focus of all evil, and their henchman, Fidel Castro, but they are also eradicating illiteracy, unemployment and providing land to the landless in Nicaragua. When the Sandinistas took over almost 80 percent of the nation was unable to read and they have reduced that figure to about 25 percent. Readers and writers have always been more threatening than illiterates.

But there are gains to be achieved from the move. When we applied a similar action against Castro in the '60s Khrushchev, then the leader of the USSR, called it a perfect move to help the Ruskies, it gave Castro no choice except to fall into our camp, the ole proletarian was quoted as saying to a Soviet defector to the USA.

Still, all this seemed somewhat less than a momentous threat to our security, or so I thought until I stumbled onto a speech about Soviet activities in Latin and Central America by James H. Michel, a US State Department Deputy Assistant Secretary. After reading his warnings I support the president's prudence all the way, those clever commies were just on the verge of taking over. Let me repeat some of his citations.

Besides the usual claims of communist support for insurgencies admittedly incited by economic and political deprivation, Michel noted that in the last

25 years the USSR had increased its number of diplomatic contacts in the area from only five to 16 countries. And they actually operate electronic intelligence facilities out of Cuba spying on the USA—how unethical can they get? They are buying a billion dollars worth of grain from Argentina every year, and (NOW HOLD ON TO YOUR SEAT FOR THIS), they provide more scholarships to Latin Americans than the US does. He did not add that if all Reagan administration proposals had been adopted by Congress the Soviet scholarship gap would have been even higher.

Michel's next paragraph after that litany of crimes was, "The nature of this varied Soviet presence. . .pose a major challenge to US interests." He went on to note that the Soviets in their dealing with the major countries of Latin America were seeking to appear "as a responsible member of the international community." Wow, do you see what I mean? These guys are insidious.

Using charts and graphs Michel went on to note that the Soviets were aiming broadcasts into Latin America that totaled more than a hundred hours a week, had 2800 men in Cuba, attempted to turn Grenada into "an unsinkable aircraft carrier" (this was quantified down to the point of noting that 3 Bulgarians were on the island), that they spread malicious rumors and distortions about our invasion of Grenada, that they sold military aircraft to Peru after we failed to do so, and that they had increased their scholarship awards from 6,320 in 1956 to over 110,000 in '82.

Well, that does it. I am mad. Mr. President, do whatever is necessary, close the scholarship gap, spy on the Russians from someplace like Turkey, tell lies about their invasion of Afghanistan, open more embassies and just do whatever it takes to get rid of those unscrupulous people who are trying to "appear as responsible members of the international community." Why if things get bad enough we might even act like one ourselves, but only as a last resort.

Lamentations
On Life's Little Irritants

12/12/82

Things that turn me off are: porno shops that decorate for Xmas, seeing the same commercial more than once during the same TV show, or how about made-for-TV commercials that run on the radio.

Or politicians and bureaucrats who are afraid to admit they made a mistake. Fanatics over anything and especially those who believe the only alternative is to destroy that which they oppose. Movies and books that make sex and profanity their central theme and thus deny the viewers and readers the fun and pleasure of experiencing their own imaginations at work.

Undershorts that crawl up my leg. Gun advocates that resist any restrictions on gun or ammunition sales and insist time and time again that guns don't kill, people do; they are calling Americans the most violent folk on this globe by their strategy of insisting that the availability of guns has no effect on our infamous murder and crime record. Flat tires, dead batteries and windshield wiper fluid containers that are empty right after a rainstorm.

On the international scene I do not like Jingoists or people who think that foreign aid means throwing money away. Most of it is used to buy US goods and employ US workers and results in a three or more multiplier effect in our economy. I am disappointed to think that our leaders are cozy with the government of South Africa,which rules by force and deprives the majority of its people their basic rights, and which deprives all of the people of nearby Namibia their freedom even though the whole world is calling for their emancipation. And I am even more turned off to learn that through our government's mislabeled "frugality," only two Namibians are able to go to college in the USA while 600 or so are studying in Cuba—the language used in Namibian schools is English. Or that more people from Latin America are receiving government sponsored educations from the Soviets than from Americans for that same "no bucks available" reason.

I'm turned off by unions that demand pay raises even when their company's profits are down and/or out, and by prices that rise even when sales are in decline. I am bummed by the number of laws Congress and the various state legislatures pass each year. It seems to me that after more than 200 years of practice, we should have long ago found most of the things that require regulation.

I'm ticked that Khomeini has turned out to be as bad or worse for his people as the Shah, although I am glad that at least this time they were able to pick their own loser. I am disturbed that after all the lives and dollars spent, life inside Vietnam has gone from bad to worse, but pleased that at least the sudden and violent death rate is down. I am angry that our nation seems to have learned little from Iran and Vietnam, and is now edging into that same kind of action in El Salvador and Nicaragua, and that our president, when asked about it, refuses to answer. This is not his country but ours, and he owes it to the people to respond to the rumors and accusations. We should not be engaged in military operations against other peoples, even by proxy or in a support role, without an act of Congress signifying approval of our citizenry.

I am also turned off by male chauvinists, even though I have clearly been guilty of being one more than once, and I especially worry about those guys who say they never have been.

I guess there are a lot of other things that tick me off, but the trouble is that when I try to think of all of them they are crowded out of my mind by all of the other things that are so good in my life and in this country. I can tolerate all of those irritants because we live in a land that provides freedom of choice—what we find wrong about it can be changed through our caring, inquiring, working, and deciding to make it better.

For example, George Shultz has gotten us off of that stupid pipeline sanction kick; Begin's excesses have driven us to a slightly more sane policy in the Middle East; and Congress, feeling the pressure from an ever increasingly concerned populace, is calling our excessive reliance on military solutions to question. Those bright spots will require more hard work before being translated into success, but my faith in the great people of this land keeps me hopeful and optimistic.

Perfect we ain't, but we are darn good and getting better. Thanks folks, and keep it up.

CHAPTER FOUR

Central America

Polls consistently show that Americans are confused and concerned about Central America. A check of the actions and rhetoric of the US government (USG) also makes it clear that it too spends a lot of time thinking about what to do in Central America.

The columns that follow suggest that the USG has let East/West issues rather than Central American realities dominate its thinking about Central America as we did in Vietnam. We are more concerned about appearing able to stop the spread of Marxism than in dealing with the poverty and powerlessness of the great masses of our Central American neighbors. I include columns on Central America in this book, because it has the greatest capacity to lead to an American diaster in the near term future. America has an almost 200 year history of mistakes in Central Ameri ca, all well documented and with excellent experts around who can recite them chapter and verse. It is indeed an area we should know about, but we are reacting to incidents down there much the same as we have to those in far away lands where our mistakes were, in retrospect, attributed to lack of knowledge.

I am convinced that if America continues to pursue the policies of the Reagan administration in Central America we will inevitably find American soldiers fighting to support the Central American oligarchies and their American corporate friends against a popular revolution of the masses seeking the American dream.

Nicaragua is probably the key nation in Central America because it is the one where a foothold has been seized by those long downtrodden masses. If the USA could assist those people to create a better society, more caring and helpful to its masses, a pattern could be set that will eventually revolutionize that region and make it a place more like we should want it to be. But if the USA cannot adjust to the Sandinista leadership, with all its shortcomings and negative aspects, and we continue to wage war against them politically, economically and even militarily, the opposite will occur. The radical forces of extremism in that region will eventually win out, and sadly they will replace one kind of a tyranny for another—perhaps even a worse one. And they will do that because they will, at least initially, feel that their drastic variation from the US economic and political example vindicates their honor and prevents their being slaves of the "norteamericano." Who wants to copy public enemy "Numero Uno?"

I am not a fan of the Sandinistas nor their counterpart Marxists in El Salvador and elsewhere south of our border. But they are revolutionary heroes who risked their lives to change an order that no American should have supported—an order that was, and often still is, as antithetical to US principles as it can be. In many ways they are far closer to the forefathers of our country who fought for its freedom than the "contras" we are now supporting in their effort to overthrow the Sandinistas, although both sides are pushing distorted views of how best to help individuals develop pride and the pursuit of happiness.

The Sandinistas do not like us and have not responded well when the US showed some tolerance of them and their policies. It is thus easy to understand how our dislike of them seems justified and how we could turn to policies of opposition to them. But our tolerance was too shallow and void of the reality that they had ample reason, far more than we had against them, to reject and dislike us. We should recognize that patience will be needed for them to overcome a hundred years of earned hatred for us, our guns, and those we supported in their domain.

America deals with governments it does not care for on a daily basis all over this globe, and will have to learn to do that in Central America as well if the aspirations we stand for are ever to apply there. It won't be easy because we are used to having our way in Central America and just assuming that our way is the best for them as well. But those days are gone forever.

It is terribly tempting to just wash our hands of the whole affair, look the other way and see what happens a decade or so from now. But that foolishly ignores the reality that much of the trouble in Central America was created and supported by our money, soldiers, guns and political support. We have to assist in cleaning it up, and we will have to do so in a way that spurns our past solution of sending in the Marines. It is time for them to settle their own affairs and then turn to us for guidance via leadership, economic support, and quiet disapproval when it is merited. But we must let them choose their own path, supporting it as long as it helps the masses, and no matter what they call it or how it hurts our investments of the past. The people of Central America have to come fist this time, not the balance sheet of US investments.

7/22/84   Who's Who In Central America?

James Mundell, a former Maryknoll priest, once said, "Some twenty years ago I traipsed through the hills and valleys of Central America. I saw poverty, rank injustice, exploitation, and the Americans doing business with the unconcerned rich. I asked why don't these people rebel at this terrible injustice. I would never stay submerged in such filth, such indecent groveling, grinding poverty."

Well, in Nicaragua they finally did. But all, including the US government, are not so pleased.

Nowhere is the dispute over the best policy for Central America (CA) more focused than in Nicaragua, where in 1979 a revolution succeeded in overthrowing a government that had been the most cooperative of all in CA with the United States. The Somoza government, so corrupt that even the pro-administration Kissinger Commission labeled it a "kleptocracy," was overthrown primarily due to the efforts of a revolutionary group called the Sandinistas.

But like so many revolutionary groups before it, the Sandinistas were better at fighting revolutions than at providing a humanistic alternative. They have postponed elections, and lost much of their earlier support from businessmen, the middle class, and the Catholic church hierarchy due to their excessive zeal in shaping the new Nicaragua to their Marxist ideas of what should be. Political prisoners number in the hundreds or more, and the Sandinistas themselves have confessed to the stupidity of their initial policy toward minorities like the Miskito Indians.

Press censorship exists in Nicaragua, although it is clearly an unusual form of the disease. *La Prensa*, the major newspaper critic of the regime, remains open and daily distributes its often censored pages, but its pre-censored stories are also readily displayed on a bulletin board in front of its offices and furnished regularly to foreign embassies without interference.

But for many Nicaraguans, and especially the majority poor, life has improved in the last five years. A massive literacy effort is producing results, and gallant efforts to improve infant mortality and health services are succeeding. Although there have been a lot of anti-business regulations installed, Nicaragua remains a mixed economy with both private enterprise, government-owned operations, and combinations thereof. The poor are getting education, jobs, land, and a greater say over their future. Most observers agree that the majority still support the revolutionary government.

Although the nation is run by a 3-person junta, the junta usually acquieses to the Council of State, a legislative body composed of representatives ELECTED by the major organizations in the nation such as the Chamber of Commerce, labor unions, and a national women's group. Although the Sandinistas hold the majority, the Council usually decides issues based on consensus rather than majority vote.

US administrations have tried to wrest the Sandinistas out of their favorable relations with Castro and the USSR by both limited aid and pressure, but the US seems to ignore the fact that the new government, and the great masses of the people in Nicaragua, are accustomed to blaming their poverty plight on the US, due to their tyrannical masters of the past. The name of their leadership group, SANDINISTA, was taken from a hero, Augusto Sandino, who was worshiped for his battle against US Marines who occupied their nation for some 20 years, and who, on their departure, left Somoza in charge. For the leaders to now consider the US a good friend/advisor and example setter is simply not logical. Such mutual respect could only be built over time.

The same can be said of the Nicaraguan support for the rebels in nearby El Salvador. To the Sandinistas and much of the Nicaraguan public, those El Salvadoran rebels are their brothers in a struggle against the military and the aristocracy that has enslaved them for a century or so. When the US price for more assistance or lessened pressure is a Nicaraguan foreswearance of aid to the revolutionaries in El Salvador, it is like asking them to sell their soul for gold. Yet, the US then uses their failure to comply as the excuse for denying aid and economic sanctions, and also to justify our support of the Contras, the anti-Sandinista armies we have unleashed out of Honduras into Nicaragua.

These Contras, who include former Somoza supporters, have attacked villages, burned bridges, destroyed oil supplies and warehouses, and even mined the harbors of Nicaragua, killing innocent people and making life harder for all. The US administration says they are our only effective leverage to make Nicaragua stop supplying arms to the rebels elsewhere in CA, and deliver on their democratic promises. But the world court and world opinion condemn the US actions as both illegal and ill advised. Even the two Latino members of the Kissinger Commission argued that the Contra program was more likely to strengthen the extremist forces in Nicaragua than to help US interests. This is borne out as the Sandinistas have used the US sponsored war to justify many of their less than egalitarian acts, calling them justified by wartime conditions.

The Nicaraguan story calls into question the US claim that the troubles in CA are inspired by the Red worldwide conspiracy. For the revolution there clearly seems to have sprung from the poverty and denial of participation in the political process. It is indigenous, caused as much by previous US mistakes as any other single reason, and was won with very little foreign assistance. The US is highly unlikely to change the Sandinista government by economic pressures or the Contra effort. Yet, the administration seems unwilling to deal with the Sandinistas, no matter how strongly it is encouraged to do so by the other nations of the region, or how often the Sandinistas offer to do so. US Nicaraguan policies are deeply intertwined with our aims in El Salvador, and in both cases the US has made its goal more the defeat of the Left than a victory of the poor and weak against the wealthy and mighty. Next week, El Salvador.

Can We Save A Savior

In 1964 the Johnson administration called El Salvador "one of the hemisphere's most stable, progressive republics." In 1984 it is awash in revolution and bloodshed and Reagan says it is the place where the life and death struggle between evil and good in our hemisphere will be won or lost. What happened?

Oddly enough many believe it was, at least in part, the Alliance for Progress, a US economic aid program, that created this apparent turn around. For the oligarchy in El Salvador, who over the years had kept their nation the least entangled with the Yankee, found that they could parley the Alliance dollars into greater profit for themselves by diverting most of it from the other 99 percent of the 4.8 million El Salvadorans. Thus their wealth soared as the expectations of the others crashed.

The latest US commitment to El Salvador is closely connected to the July 1979 victory of the Sandinistas in Nicaragua. Carter's human rights policy had prevented his siding with Somoza in Nicaragua, but after Somoza's fall, hardball US politics made it imperative that another Leftist regime not come into power in CA. Carter had a dilemma. The El Salvadoran clique of wealthy and military were among the harshest in the region. But in October of 79, either by luck or careful planning, Carter's dilemma was lessened when a group of allegedly moderate officers successfully engineered a coup and promised the kind of reforms Americans could support.

Certainly El Salvador needed change. Its overall population was rated among the fifth worse fed in the world. A mere 2 percent of the people controlled 60 percent of the land and the political process. Additionally, although already the most densely populated country in Central America, it had a birth rate increase of 3.5 percent annually. With no history of democracy or effective reform (11 meaningless elections in 28 years) the country was a time bomb waiting to be set off.

The US has now invested hundreds of millions of dollars in support of elections, land reform programs, efforts to create an independent judicial system, unions, schools, and of course a larger, better trained and equipped, professional military subservient to elected officials. It also is touting an Alliance-like economic development program.

But this five year US effort has little to show for its endeavors. The initial coup members have resigned and one of their earliest supporters now heads the rebels. The leadership of the Catholic church in El Salvador still claims the military kills more innocent people weekly than the rebels. Corruption is a daily affair within the government and military, and even the CIA asserts that the government is not winning the war against the rebels. The land reform program is touted as a sham, and the vaunted elections that put Duarte in power are looked on by many as a useless, mere cosmetic exercise, more helpful in the US than in resolving the problem. Savvy and fearful El Salvadorans know to vote for those with the guns, and the US favored candidate had the best access to guns. But Duarte had two chances before and failed miserably to turn the tide against arrogance and tyranny. Why should he fare better this time?

Supporting the government's position is the Kissinger Commission, which after a mere six days in the area paints a confusing picture; admitting to the poverty and deprivation as the major causes of the revolt, it also argues that the military must be strengthened, and that the rebels are under the spell of the Reds and thus unworthy. US intelligence reports add support as they keep trying to prove conclusively (so far unsuccessfully) that the rebels are supplied, inspired, and directed by foreigners.

But those who dispute the US policy are equally formidable including the last two presidents of Mexico, most of the elected leaders of Latin America and Europe, the most renowned US Latin American specialists such as Eldon Kenworthy, and Walter LaFeber, and former diplomats to the area like Robert White and Murat Williams. Strangely, those with the most knowledge of the region tend to align themselves against the US policy, and perhaps that is why, if only instinctively, the majority of the US public does also.

The Contadora group of four Latin American states is urging the US and Duarte, whom almost all agree is well meaning, to negotiate with the rebels, allow them a share in the government, strip the traditional military of their control over security, and hold new and truly open elections in which the rebels would be protected by an OAS police force rather than avowed foes. All other approaches, they insist, will fail and merely allow the bloodshed to go on and on. The US retorts that such would be tantamount to letting the fox into the chicken house.

They point to successes, the election's large turnout and selection of the moderate over a radical, the alleged refusal of the rebels to submit to the electorate, the declining number of death squad murders, the failure of the rebels to sustain any offensive, and the rebels increased interest in talks since US pressure has tightened against Nicaragua. They argue that although the opposition claims to be pluralistic, it is really dominated by Reds who either are or would become puppets of Cuba and the USSR. There is considerable evidence in support of this claim of communist leadership and foreign support and clearly the rebels too have committed atrocities against the innocent and their foes.

The "Savior of the World," as El Salvador is sometimes translated, has clearly become a home of seldom equaled violence for which all sides must share the blame. There is no black and white in El Slavador but many shades of greys. There are risks involved in every option and each offers plusses and minuses that will hurt innocent people. The US has chosen to follow its traditional response, as it did in the '50s in Guatemala. At that time our policy was proclaimed a victory for the American way, but today there is considerable doubt. Perhaps a look next week at Guatemala will help us better to fathom the circumstances in CA and the best ways to resolve them.

8/5/84

Guatemala, Land Of Misery

Located just south of Mexico, Guatemala is the largest by population of those states on the Central American peninsula. It is about the size of Kentucky, but as far from that state as one can imagine in prosperity, health, and security. Guatemala's people have a misleading per-capita GNP of almost $1200 per year, but a 1982 study revealed that more than half of the population, mostly the Indian half, only take in about $80 a year, that is not a misprint, ONLY $80. Another recent study suggests that the average Guatemalan's daily intake of food fulfills only a third of the minimum diet needs. All this while the Guatemalan government spends the highest percentage of its GNP on miltary in the region. Most authorities consider

Guatemala to be the most unsafe place in our hemisphere to live. There are more than 35,000 Guatemalan refugees in southern Mexico who have fled the outrages of its military.

Guatemala is rated so unsafe because in the last roughly 30 years its military has become the most violent and dangerous to its own on this globe. Between '66-'76 they killed about 50,000 mostly innocent peasants, and have been known to chain peasants together and march them off to a rich estate where they are forced to labor. In the late '70s employees of a Coca Cola plant received $2.50 for a 12 hour workday, and when they struck for a raise, their leaders were killed. Unlike most of the other military dominated societies in the region, the Guatemalan military not only fights the poor, the rebels, the reformers, and the church, it fights among its own leadership as its Mafia-like officers vie for a bigger cut of the power and wealth pie.

Much of this current picture can be traced back to numerous US interventions in Guatemalan affairs, both by troops and economic pressure. The wealthy class and most of the military in Guatemala are of Spanish descent. The poor are mostly Indian of Mayan heritage. The Spanish long ago discovered that it would be easier to get rich and stay in power if they attached themself to an established power rather than trying to convert a backward (60 percent illiteracy even today) population into a productive force. Thus they tied into US businesses, US business men, and the concept of unfettered free enterprise. But if you believe unrestricted private enterprise produces trickle down, Guatemala belies that theory. The rich prospered and the poor got poorer.

In the mid-fifties, under US reformist pressure, a real election was held and the people chose an unexpected new leader. This new JEFFE, declaring himself a proponent of Jeffersonian democracy began instituting reforms. You know the kind of things that some people say ruined the USA, allowing unions to strike, minimum wage and labor laws that limited the length of the regular work day, restricted the work of children, etc.

He also instituted land reform, carefully seizing property that long had been held out of production on the wealthiest estates. He promised to compensate the owners based on the ridiculous value they themselves had arranged with the tax authorities. Unfortunately, some of that land belonged to the largest US concern in Guatemala, United Fruit, which held 42 percent of the land in the nation. In the following months despite predictions of doom, Guatemala's economy reacted favorably to the reforms with exports up and more than a 100,000 former poor becoming landed and loving the man who made it so. The poor were even given administrative control over the land reform process, the first such opportunity for them to participate in government in 400 years.

Naturally, the oligarchy and the USA were not pleased. Arbenz, the new leader who had received rave notices in the USA upon his election, was now labeled a communist and the CIA was unleashed to arrange his overthrow. The story is best documented in a book called BITTER FRUIT, and the result was predictable. He was ousted by force and a US selected military man installed as the new leader. The reforms were cancelled.

The latest leader is General Mejia Victores. Victores, under Reagan administration pressure, has promised elections and return to civilian rule, but in the interim the army kills all who do not toe its line, and battles amongst its units for position, wealth and power. Do not hold your breath waiting for that return of democracy.

Guatemala today is a hell hole of misery and death. The very policies the last two administrations of the US have touted for dealing with the CA dilemma, were followed there in the '50s: support the educated, well to do and their military henchman, but while doing so demand elections and pressure for reform. The Guatemalans played the game and the result is not a pretty sight. And perhaps worst of all, in Guatemala today the US is a nation without respect, hated equally by rebels, military, Mayans and the rich.

8/19/84   # The Country With The Canal

In spite of all the bad news about Central America (CA) there is also some good news. Several nations in the region appear to be winning their struggle for economic and political success. Costa Rica, Panama, and nearby Colombia and Venezuela may show a better way for their neighbors to follow.

I always enjoy writing about Panama because some Yanks (fewer each year thank goodness) get so emotional about THEIR canal. Certainly one should be allowed to feel some smugness and pride about an engineering feat like that of the Panama Canal, but to me, beyond the technical achievement side of the story, the old canal ought to be listed more as an embarrassment to the US than an accomplishment.

For in order for the USA to build the canal first it had to create Panama by arranging and militarily supporting its break away from Colombia. Then we engineered a clearly fraudulent treaty giving us the canal in perpetuity for a ridiculously low price. Many historians call it so, because the man we claimed represented Panama in the treaty process was really a Frenchman, with much to personally profit from the deal, and we accepted his signature as bonafide even though we had a message from the new Panamanian leadership saying he did not represent them. One US congressman, after hearing T.R. defend himself against charges of impropriety in the case, declared that the president sounded like a man defending himself against charges of seduction by instead admitting to rape.

But nonetheless the canal project was soon underway and was eulogized in press, film, book, and memory as proof of the US's emerging greatness, a gigantic marvel attesting to the USA's way of life. But unfortunately, even that accepted wisdom is more myth than reality. When US citizens root root root for the home team, using the Panama Canal as their symbol of the majesty of private enterprise, they apparently do not realize that they are proffering a better advertisement for communism than capitalism. For the canal, great achievement that it is, was built not by private monies or companies, but by the US government with tax money. The workers, mostly black and non-US citizens, were housed, fed, medically provided for, and paid all by the US government in one of our most socialistic endeavors.

For the next many decades Panama existed as one of the most US dependent of the CA states. We had created it, we built and ran its most useful enterprise, and we occupied it and managed it in a manner befitting our alleged superiority. The canal zone became a communist-like enclave of subsidized government housing, clubs, medical facilities and whatever, sharply contrasting with the poor, but nonetheless more colorful and capitalistic Panamanian surroundings. Panama's little wealth consisted almost exclusively of the meager canal rent we doled out.

But the Panamanians finally wised up, the one factor able to rally them being a blatant anti-US sentiment that was a natural outgrowth of seeing us prosper enormously from THEIR canal. Finally General Torrijos rode to power on the promise of regaining the canal.

The 1977 treaty did not win very many votes for Jimmy Carter, but almost without exception historians and Latin American specialists, although frequently condemning the treaty as still somewhat humiliating to Panama, consider it a move that won the USA some long overdue respect in Latin America. But it did more than that for Panama. The increased revenues and higher US payments finally gave that nation a chance to establish some economic independence and potential for prosperity. And mostly it gave them a sense of respect that has enabled them to develop their own political structures and a confidence that they can do for themselves. An independence typified recently when Torrijos, just before his death, replied to a strongly worded message from the Reagan administration warning him to drop his support for the Sandinistas with words to the effect that the message must have been delivered to the wrong place, it was obviously meant for Puerto Rico, a US dependent, not Panama.

This new found sense that they can be free of US tutelage is, I believe, a requirement before CA can begin to overcome its difficult conditions. Before they can become truly independent and potentially prosperous, they will require more leaders like Torrijos, even with his many warts, so that after they develop a sense of self-reliance, they can then move on to even more useful and needed men like Belisario Betancur of Colombia, men with political skills, a sense of the need for reform, and support from large sections of their societies rather than just the rich, the military, and the Yanqui. More on Betancur and the Columbia model next week.

9/2/84 # The Dissenters Are Right

The Reagan administration asserts that the turmoil in Central America (CA) is sponsored by the USSR, and that the region has become a testing ground of responsibility and leadership between the superpowers. Thus, the

US either must assist the existing governments in CA, even with their obvious flaws, or be judged by friends and foes alike as a fainthearted failure doomed to live in a world with the Reds on the rise and us in decline. But my recent review of the situation in CA disagrees with that conclusion. The arguments of the dissenters seem to me to be more persuasive. All over the world Latin experts and other government leaders are positing that the crises in CA are homegrown, and that the roots are deep and based on situations that were created long before Lenin or Marx, before the USSR went communist or Castro turned Red.

Nonetheless, there is little doubt that the rebels, frequently, are assisted by the USSR and its allies and led by people who fall further to the left on the political spectrum than most US citizens would find comfortable. They are more socialistic economically, more central government oriented, and less enchanted with the usefulness of frequent and contested elections, diffused power, and the plurality of ideas. Thus, if they win, the resultant governments will not be pleasing to the USA.

So our current government opposes the revolutionary movements in CA while trying to modify the existing governments, to pressure them into reforms while assisting them to win the wars. To do this we spend billions, pay for attacks against sovereign nations, and sometimes even conduct those attacks. It has been interesting to observe US commentary on the recent minings in the Middle East. We have called it wanton barbarism; yet, we did the exact same thing to harbors in Nicaragua, and told the world court to go to hell when they ordered us to cease and desist.

The US seems to be supporting a double standard that is totally incoherent unless explained simply as anti-minileftist. Anything small governments or groups do or advocate that is to the left of Reagan, and almost everything other than Falwellian is, is painted as inimical to US interests and to be opposed by whatever means the Congress or public will allow. While simultaneously we carry on relations with powerful communist nations like China, and sell grain and other goods to the heart of Reddom, the USSR. Thus, it is not just Reds the US openly opposes but tiny and weak Reds.

In other words I am suggesting US policy in CA has been that of a bully. We insist that they dance to our tune, and if they do so on the surface, in other words hold elections, encourage private property and capitalism, that is good enough and we don't give a dam how rotten or perverse the system might otherwise be. But if they don't follow our rules, we step in and make them do so. We justify this as required to maintain US credibility.

But the effect seems to me to be otherwise. What the USA leadership needs to do is to send some of its top people out disguised as paupers rather than princes and observe and listen to what the great masses of humanity around this globe think of us. The vision they hold is not the one we would want or hold of ourselves. We are becoming as equally associated with evil, despotism, greed, and deprivation of human rights as the Soviets. In fact in much of the world, erroneously but nontheless so believed, the USSR is accorded the status as a greater supporter of people's rights than the US.

We could and should change that false image, but we will not if we continue the current and past policies toward CA. We must side with the great masses of deprived people, do what is necessary to strip the military and the oligarchies of their monopoly of power, and take our chances with the resulting governments, supporting them even if Reddom does also, and even if they frequently reject our advice, and take actions that we consider ill-advised. With patience we should be able to count on their inherent wisdom to finally show them that we are not only not their enemy but that our system, not the one they have had with the same name but a far more regulated and participatory one, is clearly superior to communism, socialism, and of course the totalitarian or authoritarian systems they have today.

I started this series with a quote by a former US clergyman, Charles Mundell, who proclaimed, "Thank God" when the people of Central America finally revolted against the injustice he had observed there. But if God IS on the side of the revolters, whose side are we on? —"Sorry, God!"

8/26/84

Columbia On The Right Track

Mention Colombia to most US citizens today and you will most likely either conjure up thoughts of the cocaine traffic or Juan Valdez and his exquisite coffee bean. But there is a more important story going on in that nation, for it, and Costa Rica as well, provide excellent case studies about how the problems in Central America (CA) can be dealt with successfully and more or less peacefully.

In 1982 Belisario Betancur won a convincing victory for his National Movement coalition in this land of 27 million. The election was relatively free of violence and fraud, and polls indicated that many former supporters of other parties plus numerous previous non-voters swung their votes to Betancur. His government includes, by law in reformed Colombia, "adequate and equitable" representation of the party that finished second in the election. Thus the administration has support from some 85 percent of the electorate (45 percent for him and 40 percent for the runners up).

With that kind of support, rare for Latin America, Betancur, slowly but surely, has moved to unite his nation on a path of reconciliation and increased political activity. He has been able to weaken the power of the military; he successfully retired the Defense Chief who publicly appeared lukewarm to Betancur's amnesty offer to Colombia's surviving rebels, and just recently conducted successful negotiations with the largest rebel force. Betancur is a conservative but one who recognizes how deficient Latin America has been in providing basic civil and political rights.

As a gesture to the Left he invited the Nobel laureate, Gabriel Garcia Marquez, to return to his homeland and live safely and write freely. He is an astute politician, working out deals with liberals and communists, and has moved to deal with the recession by delaying, but not stopping, the availability of no down payment homes.

In his foreign policy he has recognized that to be convincing to the disaffected in Latin America, he must show an independence of action from the dictates of the USA. He has brought his nation into the Non-Aligned Movement, and joined the Contadora group in their efforts to find a Latin solution to the difficulties in CA.

Two years prior to Belisario's election Colombia was considered to be on the verge of collapse and threatened to be pulled into the cauldron of revolution like its northern neighbors, but not today. For Betancur has recognized that national unity and success are more likely under reasonable reform than via the barrel of a gun. Colombia, Costa Rica, and Venezuela are all examples that belie the US administration's contention that wars with guerrillas are won on the battlefield first with the reforms to follow.

Belisario seems to have intuitively realized that the era of oligarchic control in Latin America is on the way out, and that if politicians rather than soldiers are to rule those lands in the future, they must unite the electorate by proving that more than just the rich or the soldiers can benefit from the political process.

If the US is wise it will give much aid and even more attention to the Betancur government and the similar efforts in Costa Rica. For they, not the militant programs afoot in Guatemala, Honduras, and El Salvador, offer the answers to the problems of CA. Vastly enlarged political and economic opportunities, respect for human rights, the down play of force, and the right to develop their own special Latin system, not a mimicry of the USA, are what the USA should be encouraging in Latin America. A review of our recent policies suggests we have not done so.

Grenada: In Which American Tradition?

10/29/83

Early in the morning of October 24 approximately 2000 US military forces and about 300 Eastern Caribbean soldiers landed on the island nation of Grenada and began to take over, fighting not only the military junta that had seized power there a few days earlier but Cuban soldiers and Soviet advisors, if they resisted.

The immediate world reaction was one of shock and criticism.

What are the opposing explanations for support or criticism?

Grenada had been cited several months ago by the US president during a speech in which he was attempting to justify the significant increases he was proposing to the US Defense Budget. He noted that the island was being run by an avowed Marxist who had close relations with the Cubans and Soviets, and that the nation was building a large runway, with Cuban assistance, that when finished would be able to handle military traffic capable of reaching targets throughout the Caribbean and even the US.

When that government was overthrown in a bloody military coup, the former Marxist president and several of his cabinet members were killed, and the group of military officers who took over sent confusing signals. They established a curfew and threatened to kill any violators during the period of the curfew. Reports out of the country indicated that some innocent bystanders as well as political resistors were shot. The US president was informed that approximately 1000 Americans were on the island, mostly either tourists, retirees living there, or students at the St. George Medical School, a school that catered to Americans who wanted a medical degree but had been unable to gain admission to the limited number of US classes. The school is a major source of medical care and income for Grenada. The president had reason to believe that at least some of the Americans were concerned for their safety and wanted out. Negotiations with Grenadan officials were ambiguous. They consistently indicated that US citizens were safe and would be permitted to leave if they wished, but who was in charge, if anyone, was questionable and it was reported that the airport was closed and that some Americans had been denied exit.

Other island nations in the region had been as suspicious as the US about the direction Grenada had been going for the last several years. Six of them urged US intervention. This request seemed to be predicated on the fact that Grenada was developing a larger military force (2-3000) than that of all of its neighbors combined. The US administration justified its decision as necessary to ensure the safety of the Americans involved (1000 out of 110,000 people on the island), because it was requested by other Caribbean states, and in order to restore democracy to Grenada. The attack was launched and is underway at the time of this writing.

In opposition the Grenadans, much of the world including almost all of Latin America and the head of the Organization of American States, have argued that the US has committed a blatant act of aggression and an illegal intervention into the affairs of a sovereign state, a clear violation of the UN and OAS Charters both of which the US not only signed but has strongly supported in the past. The British appear especially offended since Grenada is a former colony of their's and still officially a member of the British Commonwealth with the Queen as the titular sovereign. Mrs. Thatcher, a usual Reagan and US supporter, has publicly reported that she urged the US not to invade and refused to co-participate when asked by the US to do so.

Additionally the American chief executive of the medical college in Grenada reports that he advised the US not to intervene. He had concluded, after extensive discussions with the new Grenadan officials and his students and staff, that the Americans were in no grave danger, although living through a typical coup with the inherent uncertainties that evokes. He argued that an invasion might actually increase the danger for it set up hostage situations and reprisals by Grenadans that otherwise seemed unlikely.

Former ambassadors to Grenada and other islands in that Windward chain have also generally been quite critical of the action rebuffing the claim that a long runway on a tiny island is a real threat to the US and arguing that the new government, which showed signs in its few days of existence of being less cooperative with Cuba and had proclaimed a commitment to a mixed economy, deserved more time for evaluation before being overthrown by a major power swatting a mere gnat.

A final gnawing issue is that the US media has been deprived access to the region by the administration in a complete reversal of past precedent and accommodation to the idea that the US public is entitled to hear more than one view about anything it does.

The decision is yours. Has the US acted wisely and in the tradition of its democratic and self-determination principles or more in the abhorrent tradition of a hoary past of gunboat diplomacy determined to have things our own way, especially when dealing with lesser powers?

11/6/83 # Ends Versus Means

One of the oldest of philosophical debates concerns the question of which is the more important, the ends or the means. That issue is once again especially relevant to Americans as they consider the issue of Grenada and its implications about what kind of a nation we are and want to be.

It is clear that most Americans, at least for now, approve of what the president has ordered and what our military power has wrought in the Caribbean. I find myself sharing in the relief that apparently all Americans residing in Grenada have emerged unscathed from the tense situation there, and confident that the ultimate outcome is likely to result in a better life for all who reside on that tiny island. For there is little doubt in my mind that the Marxist group that had ruled that nation for the last few years, and the perhaps even more radical branch that had taken over the week before the US invasion, did not and would not serve the populace well. Thus, whatever government eventually emerges is almost bound to be, at least initially, more representative, less powerful and therefore better.

But that does not allow me the luxury of approving the means used to reach this improved goal. And I suggest those of us who use "outcome" as the means for judgement are running a great risk, for if we do so we liken ourselves to the communist totalitarians we claim to abhor and oppose. They are the masters of the concept that the end justifies the means. "For the good of the cause" is the dominant communist slogan used to justify almost any atrocity if it can be concluded that the act will ultimately contribute to the spread or success of communism.

The USA has long prided itself in the sometimes myth that we are a nation of law and not of men. That we believe laws and not the whims of leaders should govern the activities of individuals, groups, and nations, and that just laws should be adhered to even when they occasionally appear to result in less than the best conclusion. But if we allow people to break the just laws, even for the accomplishment of a particular good, we will inevitably diminish the effect of the law, open the door to repetitive violations, and allow an anarchy based on power and short term expediency to dominate.

The US invasion of Grenada violates just international and regional laws. It is an intrusion into another sovereign power's territory, uninvited by any semblance of officialdom within the raped land, and against the advice of the nation most responsible for Grenada's existence and destiny, the United Kingdom. It is in the eyes of the law at least comparable to the Soviet invasion of Afghanistan, although hopefully the presence of US forces will be of a shorter duration, and the resistance from the legal inhabitants less severe because the result will be more fortunate for and responsive to their desires.

But ends cannot be allowed to justify means if we expect to rise above the banality that exists in much of the world today. Waiting until the people of Grenada demonstrated their desire for assistance, or even until US citizens were clearly, not just by conjecture, in jeopardy, may have been harder, less dramatic and decisive looking, but it would have been the only legal way. Our glorification of this clear violation of the expected norms for nations of law may provide temporary self-gratification and pride enhancement, but in the long run it serves poorly the ideas on which we were founded. It blurs the true differences between our ideals and those of the Reds, to our detriment. And if Bernard Gwertzman is correct in his reporting that the major reason for the invasion was for the US to prove that we are not a "paper tiger," than we look even worse, for a tiger that only dares to eat little babies when there are fat and full grown enemies all around him is hardly the tiger we would all admire.

Sometimes we act like a nation of asses who love their stall no matter how infested it is with manure—just because it is home. But such does not bode well for our advancement toward the dreams of our forefathers or the promises we have made to our children and the world. Good ends attained by bad means are contaminated.

CHAPTER FIVE

 Hunger and Development

One of the major contentions of my columns has been that while the world concentrates on the East/West problem a far greater danger, the North/South problem worsens, almost unnoticed. International relations specialists have taken to referring to the problems of world development and hunger as the North/South problem in counterpoint to the long established struggle of communism versus capitalism so often referred to as the East/West schism.

It is my contention that the East/West struggle has already been won, but that the adversaries are so locked into the battle they fail to recognize that reality and the fact that they have a more dangerous problem looming in front of them. The West has already won the East/West struggle not because of our superior weaponry or our closeness to God, but because the East's approach to managing society has failed. Communism has not yet become a reality anywhere on this globe, but where it has been tried, without exception, it has failed to deliver a living standard or a human right's record that can make it attractive in competition with its rival systems of socialism or modified capitalism.

I believe that communism is only occurring today where the West, as a result of it aberrant fear of communism, has virtually driven societies into Reddom by its enmity and economic and military policies. No society would choose communism of their own free will for any fool can see that it is not the best choice. The best choice seems to be one that is an amalgamation of capitalism and socialism, a mixture I call "Capsolism" that is the most popular on our globe today and is the system practiced by the most successful societies both economically and socially. I refer to the regulated free enterprise and welfare states of Scandinavia and much of Western Europe, and to the copying of those states occurring, as possible, in China, Hungary, and Yugoslavia.

But the North/South struggle is another matter. In the late fifties during the so called Era of Rising Expectations, the US suggested to the world that if they would just follow our scheme, all could eventually approach the US living standard in this century. Few still believe that suggestion. The gap between rich and poor on much of this globe seems to be widening, and we are learning that it is impossible to transfer one society's system to another successfully. We are also learning that the effects of colonialism were far deeper than just their trimmings, that all peoples have it in them to be autocratic, vindictive and oppressive, and that it is harder to deal with global economic problems in an interrelated world economy than it is in just a single state system.

As the columns that follow suggest, the world is far from agreement on how to deal with nation building and the elimination of hunger. Even in the West, we have no agreement on what needs to be done, what priorities should prevail, the role of government and the private sector, how to begin, and what works and does not. Oh, there have been some few successes like Taiwan and South Korea, or Hong Kong and Singapore, and each of those are oft touted to support a plan or two. But each of those examples are also flawed, Taiwan and Korea by clearly undemocratic and often oppressive governments that are hardly ideal. Hong Kong and Singapore by their barely hidden corruption, slums, paternalistic leadership, and relative very small size and homogeneity that question their transferability to larger more diverse nations.

The failures have been more prevalent, often leading to military dictatorships, revolutionary movements, increased poverty and ill health. In Africa today we have a famine while the world is awash in excess grain, in Latin America we have revolutions caused by poverty and in much of the South an indebtdness to the North never ever equalled in the past.

The current answer in the USA is one of benign attention. We cannot concern ourselves with the poor of the globe until we get our own economy back in shape and our own defenses back up to snuff so that we can deal with the real threat, communism, from a position of strength. That is the position of the Reagan administration. It is read around the Third World, almost three-quarters of the world's population, as a cop out. As proof that America's prosperity comes off the sweat of the poor's brow and the bend of their tired backs.

The problem is real, some 800,000 who go to bed every night hungry, 40-50 percent unemployed or seriously underemployed, extremely low literacy rates, millions of children unschooled, poor drinking water and sanitation facilities, undernourished and diseased people to the point of retardation of up to 30-60 percent of their youth. This simply cannot continue while a third of us live in luxury and splendor, and it cannot be ignored while we squabble about spending almost a trillion dollars annually on more ways to kill one another. We ignore this problem at great risk, for those poor, sooner or later, will organize and rebel to our peril.

But since we cannot seem to agree on what to do, the North essentially is either doing nothing, or continuing to do, albeit at lower levels of expenditure, what we have done in the past and what has often served us well but the poor poorly. Carefully read the column on "Food First" based on the theories of Frances Lappe. It is my contention that she best explains the cause and the solutions. Hunger is not God wrought, she suggests, but man made by political decisions, not decisions inherent to isms but concerning power, and the unregulated drive for power must be changed if we are to resolve this issue before it engulfs us all.

Amen, Frances!!

The Failure Of Aid

There is widespread agreement that world poverty and hunger have reached crisis proportions. But although almost everyone agrees that there is a problem, there is little consensus on what to do about it. As a result, relatively speaking, little is being done, especially in any coordinated way. This column begins a series of articles in which I plan to explore the questions of world hunger and poverty.

Advocated noteworthy solutions to world poverty run the gamut from pure self-help to enormous assistance and generally fall into one of these camps: (1) The Traditional Aid View, (2) The Third World View, (3) The "Ism" Solution, (4) Overpopulation/Toughlove, and (5)The Food First Approach and its variants. I will discuss these propositions in some detail outlining their major premises and the criticisms thereof.

The Traditional Aid View has been the winner for the last several decades. It argues that the "haves" of the world are obligated by morality and prudence to assist the "have nots." It is supported by most religions which tell us that charity is noble, and it is enhanced by a practicality that tells us better off people make more attractive trading partners and neighbors.

The Marshall Plan that helped put Western Europe back on its feet after World War II represents this view's most remarkable success. However, Europe was already a developed area merely in trouble. It needed rebuilding not initial development, and it had many assets not available in most of the poor nations.

More to the point of helping those that had fallen behind for centuries are the less than unanimously acclaimed successes in places like South Korea and Taiwan. It is argued that these nations have been the recipients of large amounts of foreign aid over a long and consistent haul, and that they have moved from Third World status to Second World ranking with bright economic futures.

In the process these nations not only received billions of dollars of aid and expertise, but they developed economies patterned after the donors of this aid. They listened to the advice of the World Bank and International Monetary Fund, have conducted relatively free trade, and have not gotten caught up in welfare excesses that distracted them from surplus accumulation.

The critics of the traditional view, and there are a growing number of them, argue that the traditional assistance has not been as humanitarian as it is portrayed: that this aid was more often politically motivated than designed to help the most needy, and that it has created client states dependent on the West for their continued survival and uneven prosperity. But most of all, the critics tout as proof that something new is needed the fact that there have been far more failures than successes. As a counterpoint to Taiwan and South Korea there are the more than fifty nations of Africa who are growing poorer yearly. There are places like India where as many as 85 percent of the children are undernourished. There is Zaire, the country with the worst nutritional record in the world, yet a country that has played the game of the traditional approach by the rules. And there are nations like Haiti, Mali, Bangladesh, Pakistan, Mexico and one could go on and on with the list, whose progress is illusory and whose poor and dispirited are multiplying like rabbits.

The traditional view simply has failed, they assert, because it has been a sham. Rather than an interest in helping the needy it has been part of the East/West power struggle. The only nations receiving enough long term and consistent aid to really make progress are those that the West wanted to turn into showpieces that bettered their communist counterparts in development. And even those, along with all the rest, have had to pay a terrible price for the assistance. They have had to lose their cultural heritage and exchange it for Western ways. They have developed a leadership class educated and beholden to the West. A oligarchy of wealthy who consider their own poor countrymen more a bother than an asset, although occasionally useful as the workforce to produce the materials or products the developed states want at cheap prices. This system simply makes the rich richer and the poor poorer, they conclude.

To me this counter argument is the stronger and explains the flawed status of today. Next week the Third World's favored solution.

The Brandt Plan

4/21/85

The traditional means by which the "haves" (the North) have tried to help develop the "have nots" (the South) have been government to government aid, loans by international organizations, and some private assistance. But the result has been disappointing, so much so that a few years ago a committee of internationally known statesmen spent a couple of years exploring the problem. They published their conclusions in THE NORTH SOUTH REPORT.

Willy Brandt, the former Chancellor of West Germany, chaired the group and became its major spokesman. He and his associates recommended numerous radical changes to the policies of the past, but although the report was highly praised by many, some six years later almost none of their suggestions have been implemented.

Brandt and team first argue that foreign aid by the well off has been too miserly and inconsistent. They note that most of the aid has not gone to the most needy, but to those that for one political reason or another had something to offer the donors. They also note that such political conditions change frequently and thus the aid, with rare exceptions, has been sporadic when it needed to be long term to create any chance for positive results.

And they argue that the major international lending agencies are too strongly dominated by the North, which too often does not recognize the special problems of the South. Thus, their major remedy, and one that has been solidly ignored to date by the "haves," is the creation of a new international agency to bring assistance to the developing nations: an agency whose managerial majority would be composed of representatives from the South, not the North. This agency would be funded by a long-term and significant financial commitment from the developed nations of about one and a half percent of their GNPs for 10-30 years.

The money and the expertise that it could buy would thus be devoted to long term projects with little fear that the spigot would dry up in mid-effort due to some change in political circumstances in either the recipient or donor state. Equally importantly, of course, projects would be funded, they allege, more on the merits of the need and chances for success than the political rightness of internal policies or the profit opportunities the project might proffer an already developed donor.

The Brandt proposals are obviously more complex and detailed including requests for fewer trade barriers between North and South, a relaxation of so called austerity demands against debtor nations, and special treatment in trade for the poor including higher than market and consistent prices for their raw materials.

The logic of the NORTH SOUTH REPORT suggestions is strong—so why has it not happened? The basic answer from the North has been that the Brandt proposals came at an inconvenient time, during a budding world recession and increased East/West tensions, making even the continuation of past aid policies difficult. The US, for example, is currently ranked only 16th of 17 donors on a per-capita basis giving only about two-tenths of one percent of its GNP in economic aid to the needy, and it has trouble maintaining that level with its current deficit, trade imbalance and arms build up.

Naturally the proposals are also rejected by the leaders of the current international economic agency bureaucracy. They resent the accusations that their past actions have been politically oriented, even if considerable evidence is offered. They find allies in the leadership of many of the South nations who have often profited personally by the current system of loans, renegotiated loans, and austerity programs that seldom effect the wealthy but almost always affect the poor.

And these radical proposals are also denigrated by those, many on differing ends of the political spectrum, who oppose even the amounts being donated now as excessive, addictive to the recipient, abetting the enemy, and just poor economics. It's not more aid those people need, they argue, but less, and certainly we cannot afford more. Even though in the late 1940s the US gave about three percent of its GNP in aid via the Marshall Plan and it was an enormous success from which we still profit. But the critics reply that was to developed nations merely in trouble, who already had beaten the population monster and had proper infrastructures. It was a sound investment as well as a political move to stop communism's spread into Europe. Many of the Brandt clients on the other hand are already socialistic, overpopulated, and mismanaged beyond belief. In supporting them we only throw good money after bad.

In 1985, and greatly due to US reticence, the Brandt plan seems more unlikely than when it was unveiled. Next week we continue our search for solutions by examining the Reagan administration's apparent preferred approach.

Toughlove Or Capitalism— Can They Do It?

4/28/85

We live in a world awash with grain surpluses yet ravaged by famine. A world of hi-technology and enormous wealth countered by abject poverty, ignorance, deprivation and helplessness. And no one seems to know how to change it. In the last few weeks I have reviewed some of the potential solutions, examining first the traditional view that can claim some success, but clearly not enough, and then the Third World's proposed alternative, one that seeks ever increasing aid for the poor from the better off. A proposal that most would argue has withered on the vine because the US has opposed it.

President Reagan, as we all know, is a strong believer in private enterprise. He is convinced that it is the "isms" of the underdeveloped that have held them back. They have practiced too much government intervention, subsidies, state control, and excessive welfare, trying to give too much without concern for it being earned. And in the past we, the USA, have contributed to that failing by giving too generously and helping to cripple the will of those peoples for self-sufficiency and hard work.

His solution is a strong dose of capitalism, domestic austerity by governments and free trade internationally. When nations ask for more aid he suggests they turn not to governments but to private companies who have the know how and desire to spur their economies if they are offered conditions in which the free market can thrive. Under Reagan not only has foreign aid diminished as a percent of US GNP, but so have US contributions to the international aid agencies.

But after four years of trial the results of the Reagan administration approach have proved disappointing. Our help, tied to the austerity programs, has led to some very unhappy economic conditions in places as diverse as Brazil, Chili, Zaire and Sudan where a coup directly related to the austerity recently overthrew a man we called a staunch ally.

Reagan's ism-first approach is supported generally by the "Toughlove" theory popularized by Garrett Hardin. Hardin argues that there are two reasons to aid poor nations, morality and prudence. But that the cases are exaggerated in both areas because we do not do people a moral service by hooking them on aid, nor make the world safer in so doing. They do not love us as a result of such Christian duty but resent us even more while growing worse off rather than better. He discards the prudence theory on the grounds that the poor neither become better markets in the future based on our aid nor threaten us if we do not. They are too poor to threaten us now and will either grow poorer and less of a threat on their own, or develop best independently when they clean up their acts, get population under control and realize the aid is not coming. Hardin cites as his proof the cases of India and China. He notes that India has been a huge recipient of foreign aid, while China basically has faced its even greater development problems on its own. China has won, he asserts, adapting an approach unique perhaps to Chinese conditions but with a spirit of defiance and pride. India, on the other hand, has failed in part at least because of the crippling effects of the aid. Of course, China's communal and state planned approach contradicts Reagan's ism orientation, but it seems acceptable to Hardin's similar doctrine.

Toughlove and ism disputers counter that one can only pull themselves up by their own bootstraps if, in fact, they have boots. That the Toughlove and ism approaches are mere coverups for greed and self-indulgence, and indeed have prevailed most of the time in the last 40 years in spite of smoke screens to the contrary. That what assistance has been given has generally failed because it was seldom sufficient, almost always ism oriented to the point it enriched the capitalists at the expense of the needy, and continues to do so today.

But Toughlove's critics fault it primarily on the basis of contradictory evidence. While it is true that China seems to be dealing with its population and overall poverty better than India, it is not doing so better than Taiwan. And Taiwan received far more foreign aid than India. Additionally, Toughlove ignores the fact that many of the developed nations played a major role in creating the poverty in the Third World via colonialism. Thus, they cannot be absolved from the remedy. Additionally the price China had to pay for its solitary development was extreme both in the deprivation of rights and human suffering.

So two more solutions appear inadequate leaving us with the single issue oriented approach typified by the population controllers and another called "Food First," a far more complex, untried, and critically acclaimed solution authored by Frances Lappe.

5/5/85Too Many People?

Stephen Rosenfeld writing in the Washington Post suggests that the recent African famine offers conclusive proof that in spite of 35 years of efforts to solve the questions of poverty and hunger on our globe, we have failed and that failure is an indictment of bad advice by the Western advisors and donors to the needy.

A great many of those experts and others are firmly committed to the contention that the major cause of world hunger and underdevelopment is population growth. There are simply too many people on this globe and especially in the poorer countries. If those people would just wise up and quit having so many children they could get their economic houses in order and find a decent life for their kith and kin in a few short decades.

A strong belief in the population bomb theory goes hand in hand with the "we only hurt 'em if we help 'em" philosophy. Feed them when they are hungry and all you do is allow more to live long enough to have more kids who will be even hungrier later. We do them a favor if they die off until the survivors, the fittest and smartest, learn to practice birth control.

But the "too many people" theory also has its detractors and they have good arguments on their side. How can we have too many people if we have a grain surplus? How can we have too many people if all of them could fit in the state of Texas and live four to a house on a nearly quarter acre plot? If we are only using about 60 percent of the world's arable land? If China and its billion plus can live in a country as inhospitable and short of arable land as it does? If most nations of hungry are exporting food? And if there is more than enough total wealth to allow all to live high in comparison to their current lot, and vast potential for even more?

Is the population theory merely the rich's means of clearing their conscience about the poor? Or would getting the world population growth down at least below the GNP growth do more to alleviate the suffering of the poor than any other suggested solution?

Africa is the only continent left where population growth is consistently exceeding GNP growth, and it is no secret that Africa also harbors the world's poorest and most undernourished peoples. Billions of dollars have been spent in Africa trying to devise new seed strains, to combat deforestation, to create infrastructures and yes, even to provide basic birth control information and devices. Rafael Salas of the UN Fund for Population Activities thinks we have made progress because world wide the population rate is dropping and we have population programs today in more than 157 nations.

But still the results are disappointing, the birth rate drop is almost all the result of China's improvement and that of the already developed world. In much of the poor lands birth rates are still quite high, not because of ignorance, superstition, or uncontrolled sex drives, but because these wily peasants know that for them, individually, another child is an asset, an insurance policy for survival in old age. Other forces: politics, isms, distribution inequities, unstable markets, injustice, conspiracies, and more contrive to keep them in such a state where, incredible as it may seem, their instinctive drive to procreate is valid.

As far as I can tell, the best experts on development have concluded that population control is part of the solution to poverty and hunger but not the major step on the road. First must come improvement of the lot of the poor and with that population control will follow, with or without the modern technologies that make it easier to achieve. But it does not work the other way, a nation simply cannot, the evidence to date indicates, achieve the population limitation first and squeeze development from that gain. Many have tried and all have failed.

Supporters of population control as the second or third step rather than the first argue that even China, with its strict discipline and totalitarian control, could not master the population genie until first it began to feed and clothe—to show its multitude of poor that they could be provided for without more children. Then and only then did the people respond to the logic (for their place and circumstances) of one-child families.

And how was this Chinese miracle worked in contrast to the flawed attempts of the world's second most populous nation, India? That answer next week as I review the most exciting plan I have seen for dealing with the issue of the "haves" and "have nots." The plan of Frances Lappe, FOOD FIRST, an approach that bridges the gap between TOUGHLOVE, TRADITIONAL AID, POPULATION CONTROL, and even the THIRD WORLD ALTERNATIVE, an approach that challenges all the past assumptions, spares no sacred cows, and offers hope that, after all, it really can be done.

5/12/85

Food First Looks Good

The FOOD FIRST policy for relieving world hunger makes most of us uneasy because to follow it would require us to challenge many of our most deeply entrenched dogma. Frances Lappe, her associates at the Institute for Food and Development Policy, and a growing number of researchers like Paul Harrison and Susan George all agree that for the last forty years or so most of our efforts to eliminate hunger and spur development have been counter productive. We have increased the amount of food in the world but oddly enough, simultaneously, increased the numbers of the poor and hungry. Something is amiss, they suggest, and propose unexpected but strongly argued solutions.

Lappe rejects the population theory and asserts that there is more than enough food in the world today to feed everyone in it a nutritious and bountiful diet. Hunger is caused by politics and thus cannot be solved by more aid, improved science and technology, increased trade, or anything less than a reshaping of the political/economic order that moves control over agricultural resources to a wider portion of society, especially to those who tend the fields.

Lappe in her books and lectures gives strong evidence to support her case that it is not too many people or too poor soil, too little technology, or the wrong "ism" that creates hunger. She notes, for example, that some of the least populous nations have the greatest hunger problems, and that hunger has often been exacerbated by the arrival on the scene of new technologies such as miracle seeds that make each acre more plentiful. Zaire has a lower population per cultivated acre ratio than France, yet the French have a food surplus while Zaire's fewer people have the worst nutrition in the world.

Lappe uses a comparison of India and China to illustrate her second point. In India control of the land is held by a small minority as it is in most of the world (In 83 countries recent studies show 5 percent of the landowners own 75 percent of the land). Thus when super seed grains were brought to India, the landowners discovered that they could make a profit off of lands that formerly were considered sub-standard and thus relegated to the poor. The poor had been able to feed themselves off this land even if in constant debt to the landowner. But now the owners kicked the poor off the land, with the aid of government grants made available by international aid agencies imported modern farm equipment to reduce their labor needs, and began growing the miracle seeds on huge holdings. But did this new food production nurture the poor? No, the landowner quickly learned there was more profit to be made if he exported the product and earned foreign currency to help pay off the debt acquired to purchase the technology from abroad.

Thus, the poor lost land that formerly provided them subsistence and work while the rich got richer, and more food flowed more cheaply to the wealthiest. China, on the other hand, redistributed her lands to the poor, placing them in charge of collectives and now even small family groups, providing incentives to grow food for their own and urban consumption. Their farms are labor intensive rather than technology intensive and have increased their per-acre output considerably. Lappe points out it is not the ism of China that enabled it to eliminate hunger, but that it moved power and control over land from the hands of the distant landowners into the hands of the formerly economically disenfranchised. China's food first approach abstains from exporting food if their own are unfed. Lappe is a strong advocate of family farms and is fearful of US trends away from that proven successful system. But she stresses that there is no one model for the world other than dispersing power away from the few to the many, be it under capitalism, socialism or whatever.

Drawing on the strong evidence that none of the current practices have produced much success in reducing hunger, Lappe suggests the following. The US should cease all government to government foreign aid, including USAID projects, except where the recipient government shows evidence of democratizing control over food producing resources. All aid should be in the form of grants, because loans force them to concentrate on raising the money for payback which means export requirements that most often work against helping their own poor. Give food aid only in absolute emergencies for otherwise it undermines local agriculture. Use US clout in the international aid agencies to change their practices of supporting governments that actively restrict power to the few and their push for export growth and austerity measures—these always impact negatively against the poor.

All of these moves are designed to bring about a wrenching change in the way most Third World nations are governed. Without such changes Lappe believes hunger will continue amidst plenty. Before reading Lappe I never would have assumed that trade, technology and science, monetary assistance, and foreign investment, because they almost always help entrench the political and economic elite, could do more to increase poverty than relieve it. But now I BELIEVE. Consider it!!

 # Back Off From Aid

"The best way to solve the population and poverty problems of the South is for the North to stay out of it and let nature take its course." I was especially disappointed recently when a young college student espoused that simplistic solution to the growing North/South chasm. The term "North/South" no longer refers to shooting wars between the factions in Vietnam or Korea, but to what most international observers consider to be the most important international problem, the growing gap between the rich industrial nations (the North) and the poor, underdeveloped ones (the South).

A few facts are worthy of special note. It is anticipated that 30 million children in the South will die this year from hunger. That is equal to setting off a Hiroshima-type A-bomb every other day but targeted only against the children of the South. Today there are 800 million people in the South classified as absolutely destitute. Most demographers, having studied the evidence exhaustively, have concluded that nature alone will not prevent the main cause of this excessive population growth, or even thin it out significantly. Pestilence, famine, or natural holocaust be damned—the world's population has consistently increased.

Why don't the people in the South just wise up and quit having all of those babies? Because the marginal value of each additional child to the individual family is more a plus than a minus. For example, a recent report reveals that the average family in India needs one member to devote 200-300 days annually just to gather sufficient wood supply for fuel. Such worldwide wood gathering in itself, by the way, is threatening to denude the earth and thus, possibly, injure us all.

My aforementioned disappointment springs forth partly because the belief that the South problem is theirs alone, reflects a failure by our leaders, political, educational, and familial, to instill in our youth the basic human values which I believe our nation has long supported. Any sign of such failure is to be mourned, but in this case the attitude is even sadder because the, "it's not our concern" believers are also wrong for very pragmatic reasons. The South's despair daily effects us in the North, and as their misery deepens a ripple effect extends to us. The current remarkable US living standard owes much to the South, for their raw materials have supplied many of the needs for the technology by which we have drastically altered nature's state in the North and turned it into a near land of milk and honey. It is a commonly accepted fact that we also need to sell our products made from those raw materials back to the South in order to prosper, but poor and starving people are not big consumers.

Additionally, our free enterprise democracy thrives best in a world of peace, but peace does not flourish where hatred and envy abounds. Iran, El Salvador, and Nicaragua are merely the most recent signs of what this world has in store if we continue to ignore the desperation of the South and treat it only as a part of the East/West struggle.

It should be noted that it is not the South's poor, who live so close to nature in the raw and hate it, but only some of us who live affluent and comfortable lives in a vastly altered state from nature's who suggest that nature should be allowed to take its course. Such people are not bothered by the constant pain of hunger in our stomachs, but rather from a form of feared pain to our pocketbooks that arises when others talk about assisting the "le miserables." With our living standard threatened by inflation, we are eager to accept the lie offered by big government that is the real culprit in our diminished future, that our plight has been caused by foreigners who drain our resources by their laziness and crazed sex drives. That mistaken belief makes it easier for us to develop the absurd idea that helping the unfortunate will hurt both us and them.

I suggest a more useful alternative would be to consider investing in the South, not timidly and ineffectively as now, but aggressively as we did in Europe via the Marshall Plan. To apply our wealth and technology unabashedly at something like the three percent of our GNP of those days rather than the somewhat token one quarter of one percent we now supply, and even that only with political strings attached. A nation that set as its goal the saving of even 15 million of those starving children, and that was able to offer a significant number of those soon to be a billion destitute some hope for a productive life and a reason to give up reproductive excess, would be a nation that truly deserves the title of "Great." It is easy to be considered great in terms of the power and numbers of your weapons, but much more difficult to be so in terms of wisdom. America is in a rare position in history for we have a chance to be great in both ways—the choice is ours but the answer is in doubt.

The US spends 160 billion dollars a year on military power, although we already possess an arsenal of destructive power unmatched in history. That power seems to fail us routinely; yet our solution to each failure is to seek more weapons. The 65 billion estimate for the MX shell game could feed and educate many a child of poverty and in the long run, I suggest, that would be a far better investment providing even better security. Dare to be great, in its finest definition, America.

CHAPTER SIX

Middle East/Terrorism

Most experts on international issues assert that the Middle East is the top hot spot on the globe. Between the Arab/Israel problem, the Iran/Iraq war, the rise of Islam, Lebanon, oil supplies, Soviet/US competition, Arab rivalries, and the genesis of much of the world's terrorism it is difficult to pick which of the many insecurities of the Middle East is the most explosive.

The Middle East is also another example of a place where the US seeks to wield significant influence without apparently knowing very much about the forces at work, the history and the culture of the peoples involved. It is hard not to be critical of US policies in the Middle East. Most authorities agree that at first US interests in the area were left to our own oil companies to dictate. Later we let our sympathy and guilt feelings about world Jewry dominate our thinking, and most recently we have let US/Soviet rivalries drive our middle eastern policy. This hodge podge of direction has resulted in the clear cut situation where the USA has lost influence and has become one of the most hated nations in the world to the great masses of people in the region. We are looked on primarily as either the latest in a long line of imperialists more interested in our own good than the region's, or as Israel's major supporter and the power behind Israel's defiant and aggressive actions toward her larger but weaker neighbors.

I have grouped the subject of terrorism with the area of the Middle East because the truth is that most of the so called terrorist acts in the world either occur in the Middle East or are sponsored by those in the Middle East. Note that I used the words "so called." I did so for I believe that if we are ever going to play a long term constructive role in solving the major problems of the Middle East or world terrorism, we will have to redefine our definition of terrorism. For most of what we call terrorism is not created by small bands of radicals operating out of the main stream of their cultures, but by popularly supported freedom fighters, organized, trained and the equals of America's early Minutemen and Tea Party protestors in the eyes of most of their fellow citizens. Large numbers of peoples, many in authority, view many of the acts we label as terrorist as instead the war of the guerrilla, the flea, against the elephant. Their only way to wage war against those from the West who have ruled them from afar for too long. Those we call terrorist they label as war heroes. They do not compare the deaths and deprivations of their acts to the peaceful routine, we consider the proper norm, but to the explosive conditions they live under routinely, invasions from Israel, air raids, the death of women and children from US Marines and off shore battleships firing giant guns seemingly willy nilly into their homes.

The first column in this chapter is a new one elaborating on the points noted here. It was generated by the TWA hijacking in June '85. It suggests that until the US gets over its biases and rhetoric that reveal a one-sided image of the Middle East we will be helpless in resolving the many crises there. The USA must structure a constructive policy toward the Middle East, one predicated on the belief that we must realize how important it is to all the Middle Eastern nations, including Israel, that they take charge of their own fate and not be dictated to by foreigners as they have for the last couple of centuries. Nationalism is as powerful a force in the region as the anti-Israeli feelings and must be recognized if we are to avoid a catastrophic mistake that might lead to a superpower confrontation.

The Middle East may be one of those international problems that is nearly insolvable. It is hard to admit but there are a lot of such cases on this globe. But recognizing that it can not be resolved is not the same as concluding that nothing can be done to dampen the dangers. That can be done while we wait for time to make more permanent solutions possible. Only a madman, in my view, would pretend he has a perfect solution for Lebanon today, but crazier men would be those who decide, as at times it seems our current administration has, that we should just ignore the problem and let it work itself out. For in Lebanon lie the seeds of a great holocaust, unless men of peace are willing to strive hard to understand one another and take risks to move those troubled people on a road back to a better life. It can be done but only with a vision of the future that meets the reality of today accompanied by a strong knowledge of the past.

1985
Who's A Terrorist?

"Barbarians, Assassins, Thugs, Criminals," those are just a few of the printable titles our president has used to refer to those who hijack airliners, seize hostages, bomb embassies, blow up barracks, and shoot at American citizens. His choice of words pleases many of us for in our helplessness and feelings of insecurity it is natural to rage at and denigrate those who seem to be picking on and trying to humiliate our kind. And since they are often unknown nobodies rather than uniformed warriors, it is easier to brand them with vile names than to hate their nationality or country.

But is this an accurate, fair, or even reasonable way to deal with the problem? Ah, that is another issue. Thugs, et. al. deserve nothing better than opprobrium, thus it is reasonable and indeed proper for those they injure to strike back against them in self-defense. But are we sure the current wave of terrorism springs from thuggery? A more thoughtful exploration of the international problem suggests it may be more complicated and far less black and white, good guy/bad guy. Let us take the most recent TWA aircraft hijacking as a case to explore. Innocent American civilians were deprived of their freedom, occasionally terrorized and clearly exploited while one serviceman was killed ruthlessly and another threatened. If that were the total picture it clearly would be uncomplicated featuring only innocent Americans and barbaric hijackers.

But there is more to it than that. The other side has a view too. They do not feel that they hijacked people to rob them of their coin, a la thugs, or to rape, maim and torture them for the fun of it, a la barbarians. Their view is that they struck back against the US for its support of Israel, Israel's invasion of Lebanon, and atrocities against innocent Arab peoples, including those driven from their homes in Palestine decades ago, and the more recent attacks on refugee camps where massacres of women and children took place with apparent Israeli indulgence. They also felt they were matching the aggressiveness of the US Marines who entered Lebanon under the neutralist guise of being peace keepers, but then supported a puppet government chosen by foreigners to deprive Muslims of their rights and power. And as retaliation against a US Marine force that fired its powerful weapons willy nilly into Arab villages killing and maiming innocent women and children. Our president called the mining of waters around the Persian gulf barbaric acts, yet when we did the same to the ports of Nicaragua, it was labeled as legitimate means of international influence.

To many an Arab such so-called terrorists are heroes, freedom fighters, and gallant warriors fighting a holy battle against barbaric intruders. In the specific act that set off the TWA heist, for example, they were clearly taking hostages to hold while they demanded the release of more than 700 of their brothers who had been seized by the withdrawing Israeli army, and carted off to imprisonment in Israel without trial, hostages taken by an enemy to deter others from resisting an invasion and occupation. Years ago the USA felt justified to bomb North Vietnam when some of its bases in South Vietnam had been attacked and some of its men held prisoner, and we scoffed at the Vietnamese assertion that our bombers committed barbaric acts. But those bomb raids killed far more innocent Vietnamese than the Shiite "terrorists" actions against innocent Yankees have tallied.

Indeed more Arab women and children have become casualties from Marine armor or US bombers and Israelis armed with US weapons than the total of Americans taken hostage, killed, wounded or released physically unscathed. All of this is not said to justify the acts of hijacking, bombing, or kidnapping, but in an effort to put it into perspective and suggest that all this mayhem will best be lessened and cured not by rhetoric that always brands the other side as the only villain, but by quiet efforts at understanding the mutual grievances of the parties involved. Wearing a uniform does not make killing of innocents legal or proper, nor does the lack of such clothing brand a resistor as a brigand. The former civilian clothed and British labeled "terrorist," Menachem Begin, is an honored hero in his land and a guest of presidents in ours.

The truly important point is that since world wide acts of bombing and hijacking make for a very unsafe world, we should all be interested in arresting the spread of the disease, not just in placing blame.

Most of what I read and hear suggests that the best cure is some sort of aggressive reaction, "Retaliation is all those thugs understand," we are told. Or "We should copy the Israelis, their retaliatory policy is what works." But I find that absurd. One only has to pay a bit of attention to reality to see that Israel's reliance on the sword to solve her problems is an absolute failure. She is still a nation alright, but she is in one "mell of a hess." She is almost friendless, her economy is in a monster inflationary spiral, her people are leaving faster than new ones arrive, and the nation is so split it requires a coalition of opposites to govern it. Top that off with the fact that on a percentage basis far more Israelis have surely fallen victim to terrorist type activity than any others on this globe, and it seems ridiculous to conclude that the Israelis have the best solution.

In my opinion it is far more obvious that violence begats violence, not peace. Retaliatory raids are exactly what has brought US victims to this state of affairs. If we want to clear up terrorism it seems far more reasonable to me to make it harder on the terrorists by improved security, realizing that he will always be able to crack that security system somehow and some way.

But more importantly we must try to eliminate the causes of the terrorism. And the causes of the most blatant attacks on US citizens these days are obvious, the fate of the Arab world vis-a-vis Israel, US and European efforts to dominate the Middle East and prevent its success at self-determination, and US shortcomings as it supports the powerful few in Central America against the powerless many. The best way to deal with terrorism is to clean up our act and start helping the powerless acquire control over their lives, search for fair solutions to Palestine and Jerusalem, and contain and limit Israeli use of force as well as Arab violence.

In the interim terrorist acts will undoubtedly recur. And the way to deal with them is the way we acted in the TWA incident, with quiet diplomacy and restraint, not with guns and retaliation. More people will be spared by that response than by fruitless macho acts of revenge and by incessant name calling that makes resolution harder, even as it makes us feel better. Such a policy will only be possible for an American leader if he takes the time to explain to the people that the terrorists are not, after all, in most cases, thugs, assassins et. al., but unhappy warriors fighting to right a perceived wrong in the only way open to them. And to try to convince them, the actors, the victims, and the people on all sides that we are working diligently to relieve the grievances as equitably as possible.

If we do that all such incidents will not disappear, for of course, there are those who are aberrants and those whose grievances are misguided. But we can bring the numbers down to a controllable reality, if we subsume those many that are, to some degree, warranted reactions to our thoughtless actions. The whole curative process must begin with an objective evaluation of the causes—something most nations, including ours, more often than not, are not inclined to do.

The Middle Eastern Muddle

While Americans mourn the death of Anwar Sadat, Lebanese, Palestinians and Syrians cheer with glee. In Iran legislators shout "Death to the USA," while Secretary Haig arrives in Cairo with the promise to speed up arms deliveries as if war or an uprising were imminent. Our president works overtime arm-twisting and even bribing and warning senators with the favors he can or will not grant depending on their vote that will or will not block his decision to sell AWACS and other sophisticated military weapons to Saudi Arabia. Iran and Iraq are still at war, a war which still occasionally spreads into other lands like Kuwait, thus increasing the threat to the continued flow of that precious black gold to the needy West.

To far too many of us the Middle East seems to be a muddle—a place inhabited mainly by people we cannot understand. People we sometimes label as radical, but more recently our vice-president stripped away the diplomacy and called them what so many of us say all the time, "madmen, butchers and crazies."

What really is going on there? Why and what should we be doing about it? The right answers to these questions might prevent the next world war while the wrong ones would seem to ensure it occurs.

In my opinion three fundamental issues dominate the thinking of those born and bred in the Middle East. First, Pan-Arabism, the desire to throw off the chains of the recent past and create societies based on their own beliefs and cultures, undominated by foreigners and foreign ideas be they Britt, Turk, Russian, Yank or other.

Second, the Arab-Israel dispute. Sadat's ability to make peace with Israel hinged on his ability to restore Arab honor via a settlement that dealt with the Palestinian question and Jerusalem. Such an acceptable solution must, as a bare minimum, provide for a sovereign homeland for the displaced Palestinians, and wrest at least a part of Islam's second most holy city, Jerusalem, from exclusive Jewish control. The fact that each day the likelihood of that outcome, as a result of Sadat's negotiations with the Israelis, grew dimmer and dimmer thus sealed Sadat's fate and made his economic shortfalls and power excesses less tolerable to his nation.

Third, the battle for dominance in the region between the indigenous states. All Arab states and Iran have sought to gain ascendancy in the area so that they could win out over the ancient rivalries and disputes about boundaries, religious sectarian differences, and tribal hatreds that abound and for years were encouraged and exacerbated by the Western overlords.

Gamal Nassar left a heritage that has also played an important role in the region as each successive leader from Khomeini to Khadafy to Saddam has tried to pick up the Nassar mantle and become the leader of an Arabic-Islamic world. Each has tried to accomplish that goal by either acquiring excessive military weaponry, or by gambling on dramatic changes to the economic or social fabric to produce economic or technological success

(Socialism or a nuclear reactor), or both. Most such aspirant leaders quickly learned that a bonus could be gained in their drive for dominance by frequently tweaking, or claiming to have tweaked, the whiskers of the most important of the foreign powers, usually number one, the USA, or number two, the USSR.

With these three major factors in mind the conditions that prevail in the Middle East seem more logical after all, even if they still seem confusing to our frame of reference. Americans are just not used to thinking of themselves as unwelcome intruders.

So why is the USA so interested in the region? Three reasons appear most significant. Our major impetus has been to prevent Soviet expansion into the region. Since the Middle East is far closer to the USSR than us, we considered it a likely prime target for the Red spread. Additionally we knew that the Russians and the British had often countered each other's efforts to assert control over the important water and land routes that the Middle East encompasses, and we felt we needed to replace that British role.

The second and major reason why in today's less geopolitical world we are concerned about the region is OIL. Saudi Arabia, Kuwait, Iran, Iraq and the United Arab Emirates combine to produce 80 percent or more of the West's most easily acquired oil supplies. The fear of that oil pipeline being controlled by the USSR or by so called radical Arab states who might shut off the spigot or even more drastically increase the cost dominates the thinking of many of our leaders.

And finally our interest and investment in Israel's survival has increased our involvement in the arena. We look upon Israel as a special responsibility, and rightly so, for America was the key factor in the creation of that nation in its hostile environment. Israel is indeed the only Western-like state in the region, a people and a government who do not seem so strange to us and with whom we feel comfortable and comforting.

7/12/81

Israel In Montana

Why not move the Jews to Montana? The question is almost as bizarre as the real story of how the Middle East got into its current mess.

Historically, the Jewish people have been one of, if not the most, persecuted peoples on earth. That mistreatment reached its zenith when Hitler killed millions of Jews in his gas chambers.

The rest of the world might have prevented that tragedy, and indeed the US called for a pre-WW II conference for that very purpose, but although the chamber rang with lofty words, substantive proposals were almost non-existent and Hitler was given an unspoken but obvious blank check to proceed.

After the war when the obscenity of Hitler's atrocities rocked the world, a lot of richly deserved guilt complexes were created, and this guilt reaction was parlayed by the world Zionist movement, a relatively small Jewish group

102

seeking a homeland in Palestine, into a UN resolution strongly supported by the US to divide the British Palestinian Mandate into two areas, a Jewish and an Arab, as autonomous parts of a single Palestinian state.

The US support came under strong pressure from the American Jewish lobby and ignored what President Wilson had been told almost thirty years earlier when the King-Crane Commission, after a visit to Palestine investigating the same Zionist appeal, strongly recommended against the creation of a Jewish state there. It is clearly not wanted by the vast majority who live there, they pointed out, and will violate human rights and inevitably lead to war.

They were correct. But the UN voted 33-10 to create a Jewish state, even though the US and UK, who both voted in favor, later appealed and asked that the decision be postponed. The UK withdrew their forces early and absolved themselves of any responsibility for what was to follow. After a short time the Arab armies struck, but the Jewish forces, manned primarily by men and women who had served in European armies in WW II and armed with far superior equipment furnished or stolen from the West, drove back the bickering and ill-led Arabs.

The fleeing Arab soldiers were joined in their flight by about a million Arab civilians who, rightly or wrongly, felt escape would be better than domination by their Jewish conquerors. Those refugees have never been allowed to return. Note that no vote was taken in old Palestine, and that those who made the decision to create this new Jewish state were those who would be exporting the Jews, not those who would receive them.

No Arab state voted for the resolution, yet it was in their part of the world where the Jewish state was to arise. Three more wars followed in '56, '67 and '73, each expanded the territory under Israeli control until it was more than 100 percent larger than the autonomous region given to it by the UN. These expansions brought more than a million Arabs under Israeli control. There was no place left in the Palestinian Mandate for the Arab state equally blessed by the UN.

The refugee problem and the occupied lands have become the keystones of the Arab/Israeli relationship. The Jews assert that the other Arab states could and should easily resettle the displaced Palestinians and ease their suffering and concern. The Arab leaders counter that the refugees do not want resettlement but a homeland returned. Something they add, surely the Jews, of all people, should understand.

The last war in 1973 produced ominous signs from the Israeli view. Because for the first time the Arab armies performed well and the Israeli losses were astounding for such a small nation to absorb. The message was clear to some; three million cannot forever hope to defeat 100 million.

Thus, the door was opened for Begin to stride the path to Camp David and the peace accord with Egypt. That accord returned the Sinai, the largest but most sparsely populated of the Israeli occupied lands, to Egypt, and the agreement called for more talks to resolve the problems in the more densely Arab populated Gaza Strip and West Bank. But within days of the signing of that accord bad vibes began emanating out of Israel. Begin continued to refer to those Arab majority areas as part of the "promised land" and authorized more Jewish settlements in them.

There are other critical issues fueling the instability in the Middle East, although I would not include, as Israeli officials prefer to do, the USSR's "fishing in troubled waters."

Jerusalem is a holy city to all three of the major religions that sprang from this region. The UN plan called for Jerusalem to become an international holy city, but Jewish guns now control all of the city and almost every Israeli leader and citizen vows regularly that the question of Jerusalem's sovereignty is non-negotiable. They have named it the capitol of their nation.

The Jews apparently learned all of the wrong lessons from those who dominated them for centuries, although one can hardly expect less.

But if we are ever to resolve this problem a way must be found for Israel to become more acceptable to her neighbors, and guns and air raids do not contribute to that eventuality. Conversely, Israel's neighbors must reconcile their future with the fact that Israel is there, even if they were not consulted, and that she must be allowed to exist there in peace. The USA, as the most responsible nation in Israel's creation (the UN vote was in doubt until Truman announced that he would instruct the US ambassador to vote for it) and her major current supporter, must nudge the Jewish state in the direction of accepting a homeland for the Palestinians and the internationalization of Jerusalem. For those are the bottom line events that must occur before peace in the Middle East has any chance.

In exchange for those moves the Arab nations should guarantee the security of Israel. If that does not happen we will likely never know peace in the Middle East again and the US and the rest of the world will more and more personally feel the results of terrorism and instability.

Syria: Radical Or Realist?

1/16/83

It's popular in the West to classify Syria as one of the radicals of the Middle East. The extent of a nation's radicality seems to be predicted not, as one would expect, on how the leaders are chosen or how they act toward their people but more on their foreign policy. If they have good relations with the USSR, they are automatically radical. Additionally if they stridently oppose Israel and its expansion they qualify for that title. Thus Syria is a radical Arab state.

Syria has been ruled since 1970 by Hafez al-Assad. He took over in a military coup and has ruled since in the name of the Ba'ath Party. His foes accuse the Ba'ath of just being a cover for rule by a religious minority, the Alawite sect of the Shi'ite branch of Islam. The Alawites compose only about 11 percent of Syria's population. The majority of the folk there, about 70 percent, are Sunni while another 13 percent are Christian and the remainder mostly Shi'ite of various sects. Khomeini's movement is Shi'ite.

Assad holds on to power tenuously. There have been numerous attempted coups and assassinations against him especially by a group called the Moslem Brotherhood which seeks to regain power for the Moslem majority and to establish a religious based government.

Assad wants to stake a claim as Nassar's successor by being the leading exponent of Pan-Arabia. He refers to Syria as part of an Arab Empire and has on numerous occasions attempted to unify his nation with other Arab lands.

Before Syria gained its last independence from the French the territory included either all or parts of what is today known as Lebanon, Israel and the occupied lands, and Jordan. It was France and Great Britain, who in their great wisdom, decided where the borders should be drawn, and many Syrians believe that through that act those foreigners reduced Syria's ability to be a major force in the Arab world and did so deliberately.

Syria has some fine arable land and was once the granary of the eastern part of the Roman Empire. As recently as 1977 about half of the Syrian population was engaged in agriculture but they have also developed a large oil refinery enterprise. Syria has relatively little oil of its own to produce. Its economy, in spite of many promises by Assad, is in trouble and depends for its survival today on financial support from richer lands like Saudi Arabia and the USSR.

Assad nationalized most of the nation's largest industries when he took over but recently he has encouraged private enterprise and it is on the rise. He has been tough on communists in his own land while courting the USSR for its armament and economic support.

In 1976 when the Lebanese Civil War was about out of hand Assad sent a 20,000 man force into the nation as a peacekeeper. His action was endorsed, after the fact, by the Arab League, and his troops did reduce the slaughter. He so acted apparently because he feared that a fragmentation of Lebanon would lead to an Israeli seizure of southern Lebanon, and Lebanon's fragmentation might undermine the concept of a secular state in the region.

The Carter administration attempted to reopen US/Syrian relations in spite of Assad's questionable control and civil rights support, but Reagan has closed that door. Syria will be a key to the settlement in Lebanon. Their troops are not likely to leave Lebanon unless Israel's do as well or unless the Israelis run them out as they clearly have the power to do.

Syria has complained to the Russians about the inferiority of the weapons they furnished them compared to the US weapons of the Israeli. The Russians reply it is not the weapons but the soldier. Recently the Soviets sent some experts to Syria to try and upgrade their training. It now appears they are also preparing a place for the installation of the hottest ground to air missile in the Soviet inventory. That bodes very poorly for the future tranquility of the region since the Israeli are not likely to let that installation occur without a response, and as we all know Israel ignores borders when it foresees any threat to its military control of the region.

Syria is not really radical but a typical nation in the Middle East. It is strongly supported by the other Arab states in its demand for return of the Golan Heights territory lost in battle to the Israeli, and to the need for a solution to the Palestinian problem that provides a homeland for the Palestinians not in other Arab lands, but at least in some part of the old British Palestinian Mandate. Syria's position should not be ignored if peace is to come to the Middle East.

Lebanon: Created By Foreigners, Doomed by Reality

1/9/83

Lebanon is a nation in name only. This war torn and divided Middle-Eastern land offers us a microcosm of the rest of the region and illustrates how difficult it will be to eradicate the instability, especially if we continue to act without considering the root causes and the inherent desires of the people who live there.

The French also stuck the Lebanese with a so called, but unwritten, National Pact, whereby the nation would choose its leaders from among the various key religious groups, reserving the presidency, the key position in the adopted government, for the Christians. This was justified on the basis of a mid 1930s census which showed a Christian majority. It was already a suspect conclusion in 1943 when it occurred, but is now clearly an unwarranted procedure since the Moslems have become a significant majority.

Lebanon acquired its current form in 1920 while existing under a French League of Nations mandate over the region now called Syria and Lebanon. Just as the British and the UN decided to divide the Palestine mandate into two nations, a Jewish and an Arab, the French also decided to divide their region. The French decision was obviously taken in an effort to provide a special homeland for those Arabs who practiced Christianity, the religion of their erstwhile masters.

Thus the conditions for today's disharmony; Syria considers all or at least part of Lebanon to be more rightfully its territory, and within the land itself the majority Moslems, naturally more supported by the Arab, mostly Moslem, world are seeking more power while dominated by a minority imposed on them, they believe, from afar.

All of this was exacerbated when the Jordanians and Israelis drove more Arab Moslems into Lebanon as refugees from battles with those victors. Israel's long stated interest in controlling the area south of the Litani River and her now two invasions of that region have additionally undermined the authority of the Lebanese central government.

We like to think that the Lebanese civil war of the mid-70s was caused only by radicals, the communists and the PLO hoping to gain more freedom of action from the weakness of the central government. But this is not so, the war occurred because Western style economic success was not enough when it meant the power of government remained in the hands of those representing the minority and out of step with the region. The war was a drive to either reassert Moslem power or divide the nation into separate Christian/Moslem states.

Strangely enough when Syria found it necessary to send thousands of troops into Lebanon in '76 they at first fought to protect the Christians, who were about to be defeated by the combined power of the Moslems and the PLO. Syria did this because they thought if the Moslems won, Lebanon would be partitioned, and that might offer an excuse for the Israelis to enter the South and convert it into another occupied territory, eventually to become a part of Israel. The Syrians also opposed partition for they feared a failure of the secular state approach operating in Lebanon would enable the Israelis to oppose that kind of a solution for Palestine. The Arabs have long supported a single Palestinian state housing both Arabs and Jews but ruled by a secular majority rather than as a religious state as today. Later the Syrians sided with the Moslems, as the Christians, with Israeli help, began to acquire more power.

Today there are more than forty groups operating armed forces in Lebanon and almost all of them have switched sides, at one time or another fighting their former allies. Each of these groups control a territory, collect their own taxes, run their own schools, have their own economic arrangements, etc. None of them will be eager to give up their current status, especially those engaged in very profitable things like the opium trade.

This will be especially true as long as the foreigners attempt to perpetuate the old system whereby the Christian groups have the most power and run the military, yet, there is every sign that is what is being pushed by the US and Israel. Geymael, the new president, is the leader of the most ruthless and powerful of the Christian groups, the Phalangist. He probably has more enemies in Lebanon among the gun wielders there than any other man. His brother was assassinated before he could take the reins of government,and his father believed Lebanon should be partitioned into a Moslem and Christian state.

Geymael appears to think differently for now because he is in charge and if he can maintain control he clearly has won a bigger prize than a Christian mini-state. The Israelis support him because he is beholden to them and thus offers the possibility of recognition by another Arab land and assured Jewish dominance over the southern regions of Lebanon.

Clearly things do not bode well for a quick and lasting peace. What the US and the Israelis would consider a victory would merely put Lebanon back into the status that led to the current unrest. Syria will be happy not to see a partition but still harbors a claim to some Lebanese land and as a self-perceived flag bearer of Pan-Arabism will not be content to see a Lebanon more closely aligned with Israel than with the Arab world.

Perhaps there will never again be a Lebanon as we perceived it. To pretend otherwise is an exercise in foolish thinking. The post WW II Lebanon that existed was not a heritage of the old Mt. Lebanon, city-state of the 17th Century. It was a Western creation imposed on a region where it was alien. Certainly it suffered from the Palestinian diaspora, the Syrian invasion, and the Israeli ones as well; but it has mostly reached its current state because of the greatest problem of all in the Middle East; the meddling of foreigners and their efforts to impose solutions there that meet their sense of nationhood rather than what is desired and has been shaped by the culture of the peoples that live there.

The realization of that truth should be the major shaper of future US Middle-East policy.

Withdraw Now

1/29/84

The situation in Beirut has gone too far, and the president is proving too stubborn. The United States has nothing to gain by leaving its military forces in Lebanon and should begin immediately to arrange their withdrawal.

I agree with the president that we cannot appear to give into terrorism, but there are two things wrong with applying that axiom to Lebanon.

First, the attackers against the US and French forces near Beirut are not terrorists. It is only we who brand them as such. They are patriots who legitimately represent wronged people and oppose the presence in their land of an armed outsider that they perceive has taken sides with one of the factions in their civil war and a foreign invader. They are as legitimate as was George Washington's revolutionary army.

Second, the US forces were sent in to perform a mission that has already been accomplished, the protection of the withdrawal of the PLO army. The peacekeeping mission was then added on, but it is not achievable by such a force. Thus, they would be withdrawn not because they are under attack, but because their initial mission has been achieved and their new one is better performed by others.

The most logical solution for the good of all involved is to replace the multilateral force with a true peacekeeping force under the banner of the United Nations. A force composed of neutrals and commanded not by the biased appearing United States and French, who are thought of as Israeli supporters, but by the UN Secretary General who would receive his orders from the Security Council. The longer Reagan waits to undertake this reasonable alternative the more difficult it will be.

America could still play an important role in the eventual settlement, indeed in my opinion, a far more effective one. The US could help supply and fund this neutral force that really could move out from the airport, separate the fighting factions and provide them all with the security they desperately need. This UN force will need sophisticated communications gear, transportation and other logistics of which the US is the best supplier.

Also by using our position and political clout in the UN the US could exert a strong influence on the eventual outcome and prevent one that might prove inimical to our or Israel's interests. But we need to accept certain realities that we have so far denied, if a UN effort is to prove to be more than just a cover for an American withdrawal and continued stalemate in Lebanon.

One of those realities is that the Israeli/Lebanese agreement reached with Geymael will have to be shelved. No Israeli military presence in Lebanon can be tolerated by any Arab state. But we can provide for Israel's security via a UN presence. Such a force was present in southern Lebanon and provided excellent security for more than a year prior to the last Israeli incursion. Additionally, we will have to accept some Syrian presence in Lebanon. Syria has a special and ancient relationship in the region, was invited in by the Lebanese, and her presence has been endorsed by the Arab League. The only way to avoid continued Syrian presence would be for Israel to cede back the Golan Heights to Syria, and that is a highly unlikely prospect.

Finally, Geymael, tainted by his Israeli/American support, will likely have to step aside, and allow the creation of a new political arrangement that recognizes the majority position of the Muslims rather than the Christians. If that cannot be made acceptable to the Lebanese Christians, by minority protections, than the dismemberment of Lebanon seems highly likely. The Christians then might have to settle for a small city state, much like Singapore or Hong Kong, located just around Beirut with the rest of the nation either joining with Syria or becoming a new Muslim state.

The president cannot see his way to such a solution because he is blinded by the chronic politicians' fear that they cannot admit to mistakes.

But he made one and would be far better off if he avoided the path of his predecessors, a la Vietnam, admitted the error and changed the policy. But he probably will not, unless of course, millions of Americans demand it. I think we should.

9/30/84 What We Do Not Know Can Hurt

The latest bombing incident in Lebanon is one more example of the Reagan administration's failures in foreign affairs. The administration had attempted to ignore the Middle East since its withdrawal of our alleged peace keeping force there several months ago, obviously hoping that the voting populace would forget its ignominious failure there if the subject was kept out of sight and out of mind. But it did not work.

Naturally, the government will react to the latest bombing of a US embassy as if it were just another of those random and ruthless acts of terrorism sponsored by the USSR over which we civilized folk can do so little other than shake our heads and realize what awful people we have to deal with. But the evidence indicates otherwise. The bombing is a direct result of a US foreign policy move relatively ignored in our nation.

It would hardly seem a coincidence that the bombing attack came just a few days after the US used its veto in the UN Security Council to defeat an otherwise unanimously supported Lebanese resolution calling on Israel to end its restrictions on civilians in Israeli occupied southern Lebanon.

The conservative and highly reputed ECONOMIST magazine reported that the Lebanese prime minister, Mr. Karami, cursed the USA when he learned of the veto, but our administration tries to tell us that US moves in Lebanon are highly respected by the Lebanese leadership. Almost all such acts of reprisal against the US are cast by the administration, and far too often by a pliant media as well, as senseless terrorism supported by only the radical few. That is occasionally true, but more often than not, in the Middle East and Latin America especially, such attacks on the US flag or US citizens are carefully planned responses to what are perceived as well calculated US moves to assist either Israel's military edge over her Arab neighbors or the oligarchies' power position over the masses in Central and South America.

To millions of people such acts are not terroristic at all, but heroic deeds of freedom fighters conducted against all odds against those who either directly oppress them or support their oppressors. When such acts kill women or children we naturally become indignant and label the attackers as less than human murderers, but the planners and executors of these attacks accept such outcomes as fitting for those who have either bombed or napalmed their women and children, or supplied the weapons for such attacks and supported them politically. I think it important that we in the USA realize that the initiators of such acts are the same kind of people who joined the US revolution and threw tea into Boston harbor. Men and women who feel wronged and are determined to do something about it.

If we think of them as people with unfulfilled needs and perceived wrongs, rather than as evil villains bent only on terror for terror's sake, we will seek solutions other than through the barrel of a gun and a contempt for their views and rights.

Obviously, Lebanon and all of the Middle East is a very complex issue, perhaps verging on insolveability, with plenty of wrong to be shared by Syrian, Iraqi, Iranian, Jordanian, Shiite, Sunni, Maronite, Druze as well as Israeli, American, Soviet, French, etc. But that complexity must be kept in mind at all times, even when loved ones are lost so seemingly unnecessarily in what appear as random acts of terrorism. At such times we must avoid the tendency to over simplify and callously blame it on "bad guys," leading to a diminishment of our patience and a lessening of our will to resolve the issues fairly for all. The embassy bombing was a tragedy certainly, but so was the incident that probably triggered it, when once again the US acted as if it knew more about how to solve another nation's problems than all the remaining 14 members of the UN Security Council.

Our disdain for Lebanon's leadership and people, as chaotic as their condition may be, and for other nations and experts with a demonstrated better understanding of the Middle Eastern milieu, came back to haunt us and is, I believe, an example of how our ignorance of others hurts us. Yes, Virginia, what you don't know can hurt!

CHAPTER SEVEN

The USA

Although the major focus of my columns has always been on international affairs, it is simply impossible to write about a nation's involvement with world issues without commenting on its domestic scene for the two subjects are interrelated and inseparable. We can isolate an international or domestic issue for discussion, but in so doing we always create a somewhat false picture because any decision a nation like the USA takes in the international arena is shaped by domestic realities and vice versa.

You want proof? Take the gasoline shortage in the USA during the early '70s. It was brought on by US policies in the Middle East, and was an effort by the OPEC nations to force us and our allies to be more evenhanded in our treatment of Arab states and Israel. That international problem created a domestic crisis as Americans began complaining about the gasoline shortage, the long lines and inconveniences, the government mismanagement of the distribution regulations, etc. Clearly, the US government had a domestic and international problem, intricately intertwined.

I was in Vienna, Austria at the time serving with the initial US delegation to the Mutual and Balanced Force Reduction Talks (MBFR). Having a chauffeur whenever I needed it, and living out of a hotel meant that the gasoline shortage was nearly irrelevant to me. But my family was back in the DC area and the letters from my wife turned that international incident into a peculiarly domestic concern.

The delegation had left Washington with the understanding that we would be abroad about 4-6 weeks to set the procedures and agenda for the main talks which would follow. But as so often happens in diplomacy, we had run into an unexpected snag and the talks dragged on and on, more or less inconclusively. By the time of the gas line incident, we had been gone more than three months and family relationships were becoming strained.

One morning while reading the overseas HERALD TRIBUNE with my breakfast I enjoyed a humorous column of Art Buchwald's that recited how wives in the DC area were taking advantage of the long lines to buy gasoline as coverups for their romantic rendezvouses. When husbands arrived only to find dinners unprepared, the wives quick and easy cover was that it had taken so long to buy gas, everything was running late. I chuckled until the afternoon mail arrived.

In it was a letter from my number one housewife. Now we had been married for quite some time by then and although neither of us felt that the bloom was over, we had long ago quit worrying about fancy perfumed stationary, handwritten notes, etc. Thus, the letter, like most from the joy of my life, was hastily type-written on the back of a letter from someone else and with a typewriter that needed a ribbon and had some faulty keys. Deciphering it was part of the fun.

My wife, in her usual newsy way, was telling me about her life without me around. She got to the issue of the gasoline shortage and proceeded to narrate that it was driving her nuts and complicating her days. She said the day before was her day to buy, and that she had waited in line for three hours and ---. Well what she had meant to type was "20 min.," the abbreviation for minutes, but in her hurry she stroked the letters "men" and went on undeterred. Had I not read the Buchwald article, I probably would have never noticed but when I discovered that it took her 3 hours and 20 men to buy gas, I took her letter, and the Buchwald article to my boss and said, "You've gotta let me go home for a while." Domestic and international problems can be hopelessly intertwined.

Often readers of my column jump on me for being too critical of the USA. But I always answer that what good is freedom of speech if one cannot use it to criticize, and in an effort to make your nation better than it is rather than just better than some. Most of the included articles are designed for just that purpose, to expose some of the shortcomings of what I agree is a pretty great nation, and to get people in that nation interested in holding it to a very high standard of conduct. I always also try to suggest a cure rather than just offer a criticism. I think the articles are self-explanatory, with the reason for their being written usually revealed in the text. The one on Little Rocks' Central High starts off the chapter, and is typical. It was generated by seeing that old movie and realizing how much progress we had made in an area where once change seemed almost impossible to hope for. But we had done it primarily because the people of conscience had prodded us ahead. I suggest such folk need to keep the heat on us all and move us forward in the area of anti-communism if we ever hope to build a more secure international environment.

There are three articles about elections and the electoral process our nation uses. I truly believe that our forefathers would be very disappointed by what we have allowed the presidency to become, and that they would urge us to return to the days when the House of Representatives was the cornerstone of democracy, and the president the administrator of law, not the creator of laws and policies, not the one official chosen by all of the people.

I also believe the political process of our nation has become completely estranged from what those great thinkers had in mind. I assume they would abhor the idea of professional lawmakers, of full time politicians that has evolved, and that they would urge us to bring more and more of our citizenry into the political process, not just as voters but as public officials. Something that will only happen if we eliminate professional politicians by limiting public office holding by anyone to no more than one term in any one office.

And the final theme of the chapter on the USA is that America has become enamored with the power theory of international relations. During our first century and a half, we spent most of our time criticizing those great international minded nations like Great Britain for playing the power game. We argued that we stood for something different. A more idealistic view perhaps but one we truly believed in. An idea that nation-state relationships should be subject to a rule of law, free trade, equality, etc. But after WW II when we became, via the fates, the greatest power on earth, we found that idealistic view challenged by the pragmatists, the realists in our midst and they have won. Today the USA practices power politics and Machiavellian machinations as much as any land on this earth. I bemoan that fact in this chapter and in the following one on ethics. In the last chapter on solutions I will suggest ways to blend the two approaches in a way that makes the current one more palatable and can put a special American twist on world affairs that might earn us the title of "great" in a way with far more meaning than just "powerful."

5/23/82 Little Rock And All Of Us

The other night while flipping channels I ended up watching most of the film "Crisis at Central High," the story of the integration of the high school in Little Rock, Ark., in 1957. Today's reality makes it difficult to recall how much of America violently resisted the integration of our schools. Governor Faubus, George Wallace, and their supporters were a popular branch of establishment America that firmly believed the very fiber of our nation would be destroyed by the close association of black and white.

It is encouraging to note that almost every school in America now routinely includes all races in near total equality, and even George Wallace seeks black support on his campaign trail. I wish that another one of our deep rooted stupidities revealed in the film would look as passe in the '80s. I refer to our fear of communism, a fear that has resisted bowing to logic far more than our latent racism. In the film the protestors, who in today's light look so stupid and unreal, often yelled epitaphs and displayed banners labeling the integration move as communist inspired and its supporters as Reds.

Over the years we have used our irrational fear of communism to resist all kinds of fictitious evils, from fluoride in the water to Social Security; yet, the popularity of the idea that communism is connected with all evil hangs tough in the heartland of America and thus in the rhetoric of much of our leadership.

It seems to me to be about time for Americans to learn and accept an unwelcome truth about communism, just as they have about racism: communism is not inherently evil and communists are not of lesser intelligence. Communism has become an acceptable, albeit less successful, economic alternative for more than a billion people. It has flaws, just as do socialism and capitalism. It seems that all of the economic "isms" need modification and controls in order to produce humane results for minorities, majorities, and individuals, and to provide the needed jobs, progress, social and economic opportunities that most people seek.

Competitively, communism would appear to rank clearly no better than third place in the economic sweepstakes. Like its major competitor, capitalism, it does not exist in pure form anywhere on this globe, but where it is claimed or identified as the major strain of economics and/or government, its level of success, especially in comparison, has been considerably disappointing.

China is experimenting with all kinds of market approaches to cure the economic malaise that communism has wrought there. The Poles are in open rebellion, brought on as much by the failure of the Red idea to provide for the material needs of the masses as for the seemingly inevitable personal freedom deprivations it proffers. Tanzania is on the verge of economic collapse due to the over-ambitious leftist model it pursued. Cuba's leader has advised the Nicaraguans not to abandon free enterprise. Even the USSR, the major exporter of the communist ideal, finds that communism, in spite of the vaunted claims of progress issued in the exciting '50s and early '60s, has fallen on hard times, gained little ground on the rightist West, and has built a society in which economic prosperity is best achieved through the black market, bribery, and the special privilege of party membership.

But in spite of that, it must be noted that the majority of people living under communist systems are materially, and in some cases even socially, better off than they were under the preceding economic and governmental systems. Few Russians would prefer the status of the masses under the Czars and the mislabeled free enterprise concept of that time that was in reality limited to the nobility. Few Cubans and even fewer Chinese would prefer to return to the Batista or warlord days. Judgement of communism should be tempered strongly by what preceded it, and in much of the world today some so called free enterprise antecedents were nothing like what you and I think of when we say capitalism.

I suggest it is time for Americans to realize that having a better system (albeit one that is also clearly flawed by a growing concentration into fewer and fewer hands) does not mean that all other systems are wrong, threatening and/or must be snuffed out. We must shed ourselves of the myth that communism threatens us no matter where it is installed, just as we have finally begun to shed the belief that blacks are inferior. Then and only then will we be able to realistically react to the currents afoot in the world. The adoption of communism in many places in this world, such as El Salvador and Guatemala, would certainly not be the best result for the people there, but it might be better for more of them than the current systems, and under no reasonable circumstance would be a threat to our nation.

But while we cling to the misbelief that communism is a cancer about to infest the world, we are too quick to turn to any cure, such as our support of dictators and our use of the CIA to bribe, misinform and intrude. Such acts make us appear foolish and unreal as the actors did on the screen portraying the way Americans acted just a few years ago when threatened by the arrival of nine young blacks at a Little Rock school.

Foolish, stupid, un-American—yes, it was all of those things, but it happened. Indeed, it is happening and is being viewed by billions around this world who look on in dismay as America, the land of the free, inexplicably supports bad guys against better ones out of an irrational fear. Let's change it!!

Freedom Means More Than A Vote

7/3/83

Happy Fourth! How fortunate we are to be part of the world's most successful democracy? But do you ever wonder just what a democracy is? It seems that almost every nation in the world claims to be one, yet even the most unsophisticated of analysts can tell you that most of them fall far short of the definition.

Many would argue that free elections are the key to democracy, but unfortunately that may or may not be so. Let's take the recent election in Great Britain as a case in point. Most of us are sure that Great Britain is a democracy and heard that Margaret Thatcher won a landslide victory, but try and explain to a Martian what a democracy is by explaining the Thatcher election. She won with 42 percent of the votes cast while her opponents polled 52 percent, yet from that vote Mrs. Thatcher's Party acquired 61 percent of the seats in Parliament, while the newest party in the election, the Liberal-Social Democratic Alliance, got just over 25 percent of the vote but only 3.5 percent of the seats. According to THE ECONOMIST the electoral system is so skewed that in Southern England from Kent to Cromwell, Labor candidates accrued more than a million votes yet garnered only one MP as the result of their efforts. Even the Conservative Party-supporting ECONOMIST suggests that is unfair and undemocratic.

Moving closer to home how many people do you think voted for George Washington, Tom Jefferson, or John Adams? Well actually none, at least not directly, for the US Constitution did not initially require a popular vote for the president. Our forefathers obviously were not enamored with the public's ability to choose wisely and intended America's wise men, somewhat like the Communists do it today, to pick the president. Thus it was not until 1824 that presidents were voted on by the public. Speaking of democratic action NO American president has ever received more than 50 percent of the ELIGIBLE voters votes, although some have gotten more than 50 percent of the votes cast, and 3 were elected with less votes than one of their opponents. And all of these variants on democracy are not old, recent polls show that more than 60 percent of Americans favor the mutual nuclear freeze but our current elected leader has chosen to reject it out of hand. That's democracy?

Getting back to our ancestral neighbor England, another surprise is that Mrs. Thatcher did rather well in reaping the votes of the working class even with a 13 percent unemployment figure. One reason for that is that the UK unemployed suffer less in the pocketbook than their counterparts in the USA. As a result of former Labor governments rule in London, British unemployed do not have a time limit on their unemployment benefits. Unemployment is another of those areas where democracies are baffling to their explainers. Almost everyone now agrees that a job is one of the major concerns of all citizens, yet the leading democracies are all running high unemployment rates while their autocratic rivals are able to boast of full, if regulated, employment. I am reminded of a quote that I think is quite profound from a book by Eric Hoffer, THE TRUE BELIEVER. He says, "Equality without freedom creates a more stable social pattern than freedom without equality." That is a direct contradiction of the beliefs of people like Mrs. Kirkpatrick, our UN Ambassador, who argues that democracies should support autocrats if they oppose communism, but must oppose communist states even if they create better economic equity. On a recent PBS TV show, THE RUSSIANS ARE HERE, several Soviet immigrants to the US exhibited a nostalgia for their less free but employed life in the USSR for that very reason. To me Hoffer's words justify the need to regulate free enterprise and for governments, even under a capitalist system, to ensure the care of its needy.

As a man who has lived in communist states, socialist states, capitalist states, under military juntas, autocrats, and elected officials, there is not the slightest doubt in my mind that the combination of reasonably regulated capitalism and elected civilian officials is tops in providing both for the economic needs and the inherent human desires for freedom. Define democracy and freedom anyway you want but such states, even if not perfect, seem to me to provide it best.

On this Independence Day I hope our current government will learn that the American democratic style is not only composed of capitalism and elections but also of freedom from want, and when the market economy alone is not able to provide without undue want it is acceptable and even may be necessary for government to do so. America is a great nation in 1983 but it will not become greater by ignoring the suffering of the poorest and least in its midst or by opposing labels. Our greatness will not be measured in history as much by the wealth of our richest as by the assistance we render our poorest, nor by our power as much as by our restraint and sagacity in its use. To me a democracy is where ALL are considered important enough to matter. By that definition the USA is a democracy, long may it be so.

<table><tr><td>12/11/83</td><td></td></tr></table>

US World Image Tarnished

America's foreign policy as viewed by much of the world, friend, foe, and neutral, is not receiving very high marks. This should be especially disconcerting to a president who campaigned on the theme that his predecessor was hurting America's image abroad by erratic and unsound reactions to world events.

The invasion of Grenada led to a resolution of censure in the UN that was prevented only by a US veto. British newspapers commented that the US not only ignored Britain's advice on that issue, involving a member of their Commonwealth, but that we lied to them about our plans. The event was equally criticized by German authorities, their chancellor indicating that he was not even consulted, but that if he had been he would have advised against the action. As a result, many European leaders have now asked just how credible the US promise about a joint decision on the use of the Euromissiles can be if their wishes are to be ignored on an issue like Grenada.

Lebanon is also the subject of many articles in the world press. America, while being praised for not backing off under attack, is also being criticized for not having thought the issue through when we first decided to send troops to Lebanon, for not following the military action up with sufficient political support, and for slowly but steadily changing its role in Lebanon from peacekeeper to advocate of Israeli and Lebanese Christian policies that are seen as antithetical to the majority of the region. It is simply not prudent or ethical for a party to pretend to be a peacekeeper, with the necessary neutrality that requires, while obviously developing a special relationship with certain involved members of the crisis. To the Middle Easterner especially and to much of the world the US looks like a bribed judge and policeman pretending to serve justice.

In Africa the US is now being labeled by most of the African and European states as the major block to South Africa's granting freedom to Namibia. The US insistence that Namibia's freedom is connected to the Cuban departure from Angola is preventing agreement and not deemed worthy or proper by many.

US policy in Asia is equally negative, our relations with the People's Republic of China are soured by our insistent policy of rearming Taiwan. Asian youth and responsible leaders are alarmed by our blind loyalty to South Korea and the Phillipines, no matter the human rights excesses of their leaders. Additionally the continued US effort to restrict Japanese exports to us and the ever more obvious pressure on that nation to spend more money on weaponry and troops, when their people oppose and fear that it may again lead to too much influence by the military, is tarnishing our image.

In Central America efforts to turn attention to negotiations rather than guns are constantly threatened by the US intimidation, as revealed in the lengthy US maneuver in Honduras and our pressure on the Contras fighting in Nicaragua to increase the level of their war against the Sandanistas. Grenada had a devastating impact in Central and South America. The administration pretends it has merely made the Cubans less aggressive and encouraged their friends to question allegiances with them. But the reality indicated in their media and from my contacts is that the governments and people in that region alike learned that US words about self-determination and equality are as meaningless as similar Soviet words applied to Eastern Europe.

Jimmy Carter was not a popular US president in Europe; yet his personal disapproval rating in Britain was never higher than 10 percent. Reagan's today is about 30 percent and British lack of confidence in America's leadership has risen from 30 percent in Carter's time to more than 45 percent today, according to a recent Gallup index.

In Reagan's campaign for the presidency he promised to return the US to its time of glory when the world looked at us with respect and confidence in our leadership. After three years all that seems to have happened is that many more than ever before fear us and worry about our ability to lead. His accomplishments seem limited to a victory over a tiny island and 500 Cuban work brigade soldiers, while his failures are littered with lost respect and the graves of young American soldiers sent off without a mission.

Why Voter Apathy?

3/25/84

With all the fanfare from the media one would guess that the American public is very interested and concerned about the 1984 election. But the facts belie that conclusion. Voter apathy is the name of the day with almost as many registered voters not voting in each of our elections as do. Thus, when a candidate wins, as they frequently do, with only slightly over half the vote, they are being supported by only about a quarter of our adult populace.

Many specialists have searched for the reasons behind this failure of the American populace to exercise its franchise on a respectable scale. We have the lowest voter turnout in the democratic free world, and their conclusions are not very encouraging.

For although they can make a marginal case for election reforms involving easier registration, better education, and even voting day holidays, they usually conclude the major failing is that most non-voters are unimpressed with the choices, consider no real differences exist between candidates and parties, and are tired of casting negative votes AGAINST the lesser of two bad choices rather than FOR someone or something.

This raises a question about the quality of presidential candidates. Why has the nation that once produced leaders like Washington, Jefferson, Madison, Monroe, and Lincoln suddenly been reduced to men like Nixon, Carter and Reagan much less Harding, Coolidge, Cleveland, Grant and their ilk?

The question bears particular examination when in the height of the '84 campaign the most memorable lines have become, "Where's the beef?," and "It's time for something new." Although Hart is clearly on to something important when he notes the yearning of our nation for a new and better way, both Carter and Reagan upset the conventional political wisdom with similarly based political campaigns in 1976 and 1980, pretending to be outsiders opposed to the D.C. trends. Does this too not reflect national dissatisfaction?

Maybe our forefathers understood this reality better than we, for they deliberately set out to avoid having us even think that we, the people, elected the president. They considered that method and rejected it for they did not want the president chosen by the fickle whim of the public, by people easily swayed by promise, charisma, and political posturing. They wanted a president who was a good administrator, manager, and above all someone of high integrity—something they knew only those people who worked with him and knew him well could judge. Thus the electoral college was born—not so

118

much to enhance the influence of the states as most Civics courses suggest, but to prevent the people or the Congress from choosing the president and leaving him beholden to either. Their plan was for a group to be called together who knew the national and state leaders and to choose from among the best and brightest, by secret ballot, the next president, and then disband, never to be a constituency to sway him.

But we, perhaps foolishly, have changed that unique system they invented, stripped the electoral college of its special value, and for all practical purposes, created a direct election of the president, albeit one the college can occasionally mar and distort.

The college was first changed when parties entered the process, took over the nomination of the candidates and determined the electors, pledging them to support a particular candidate. Almost immediately the greatness index of our presidents suffered, although we still got some thoroughbreds because candidates were still basically chosen by people who really knew them.

But in the '60s we democratized the system even more and moved significantly toward primaries allowing the public to choose the candidates as well as the winner. The caliber of presidents suffered even more, the voter participation declined and today the result is clear, Hobson's choice candidates and a disgruntled populace.

Perhaps we should reconsider what our founding fathers designed and ponder why almost all other free nations elect local representatives and let them choose from among themselves who will be the national leader, rather than entrusting that choice to those who have the least knowledge about it. Democracy needs be limited by what can reasonably be done, and it appears that 230 million people cannot choose the best in their midst to lead.

4/22/84   Selected Presidents Were Better

A few weeks ago my column suggested that by deviating from the electoral college system designed by our forefathers, our nation may have diminished the quality of our presidents. The argument was based on the fact that it is harder for all of us to be knowledgeable about the caliber of potential candidates than it would be for just 270 of us, chosen by the rest, to make an informed decision for us. I also suggested that the attributes that make a winning candidate in an election might just be the opposite of attributes that make for a good president.

The reaction to that column has caused me to pursue it further, and I discovered the following information that supports the contention that the quality of our presidents has diminished with the trend toward more direct democracy. I compared the results from five polls in which historians gave a quality rating to all of our presidents who had been in office for most of a full term. The ratings were remarkably consistent.

I then divided the presidents into three groups: those, more or less, selected by the electoral system, those elected almost directly by the people, but selected for the nomination mostly by party officials, and those, more or less, both nominated and elected by the people as the swing to party primaries got underway after World War II. None of those descriptions are perfectly accurate for all of those included, but as we go from group one to three the presidents were all elected less and less as designed by the constitutional framers.

Based on the total score possible in the five polls the candidates could record between five and thirty points with 30 being a unanimous GREAT rating, and a 5 meaning a unanimous FAILURE RATING. The first group included Washington, both Adamses, Jefferson, Madison and Monroe and attained an average quality rating of 23.33. The second group started with Jackson and ran through Eisenhower and averaged only 17.7. The third group, those most democratically elected, from Kennedy to Carter, scored an average of 14.75. Reagan was excluded since his tenure is not over.

Now obviously these ratings are skewered by time, with memories of presidents tending to treat them more favorably, but the trend still supports the contention. Of that first group of six presidents, two are consistently rated GREAT and all of the rest ABOVE AVERAGE to NEAR GREAT. In the second and much larger group of 21 presidents, still, only two are rated GREAT, and both of them were in office in extremely critical times, Lincoln and FDR. In the final five none are rated GREAT or even NEAR GREAT overall.

An additional factor is that the American public was apparently quite often dissatisfied with those it more directly chose. For although four of the six in group one were reelected and served two full terms, a 66 percent record, in group three none served out two full terms (Kennedy was assassinated), and in group 2 only 8 of 21 were reelected, an under 40 percent approval rate.

Is there a way in these modern times we could return to a system that arguably gave us better presidents more consistently? One way might be to have the electors directly elected by the people, unpledged to any candidate, indeed we might even prevent their running for the honor. Just let people go to the polls and vote for whomever in their congressional district they think could best represent them. Such electors, after signing an affidavit of non-precommitment to any candidate and never being eligible for the position again, could all then proceed to a selected site and for three weeks, with no diversions, be exposed to the presidential candidates, asking questions of them, reading position papers presented, listening to a limited number of spokespersons for each, and studying them all before casting a secret ballot for the one they conclude would make the best president. Failure to gain a majority vote would proceed, as now, to the House for resolution.

Candidates would include all eligible incumbents or if not eligible their vice-presidents, up to three nominees from each major party, selected as the party determines, but if by primary limited to four regional ones in four weeks. Other candidates who acquire a large number of signatures in over half the states could be included, and we might let both houses of Congress nominate a candidate or two if they choose to do so.

This approach would perpetuate representative democracy, lead to a more informed selection, reduce campaign time and expense, and eliminate the presidential claim to a special mandate higher than Congress's. Think it over.

Beware of Changes To Make Government Work Better

1/23/83

It seems the kid gloves are finally off and the media and political foes have decided they can safely attack the presidency. For the first two years of the Reagan administration the president's personal popularity kept that from happening, but that popularity has fallen as the unemployment rate has risen and as the failure of Reaganomics has been verified. A recent poll of top business executives shows that the president's approval rating, even by them, has fallen from over 50 percent to under 25 percent. Columnists like Anthony Lewis and Bob Kaiser have accused Reagan and his team of failure, ineptitude and a callous disdain for the nation.

But the truth is the same has happened, sooner or later, to all our post WW II presidents. The failure of either party or any philosophical stand from liberal to conservative, or in between, to consistently hold the support of the American public has led to numerous outcries from both politicians and political analysts for dramatic changes to our government. They suggest America may no longer be governable.

The facts are scary. Nixon was driven from office shortly after he received an apparent election mandate for his second term, and Johnson was forced to step aside rather than face defeat shortly after receiving the largest voter victory in our history. Only Eisenhower was able to sustain sufficient support to succeed in carrying out his two full terms.

Presidency experts like Robert Shogun in his book, "None of the Above" argue that our presidents would be treated more civilly by the public if they could carry out the programs they promise during their election campaigns. But they cannot, they suggest, because the system designed by our forefathers no longer works in this far more complex world of the 20th century. The new times, they posit, demand a system more responsive to the leadership of the one leader chosen by all the people.

Most of their suggestions require constitutional change and center around the idea of moving the US in the direction of a parliamentary system. They suggest such a system would guarantee that the Congress will either give the national leader what he promised to deliver or allow him to force a new election. Alternatively they suggest we should outlaw split ticket voting and have all, or at least most, of Congress selected on a slate with the president. These and other proposals would clearly all succeed in making presidents more able to work their will on the nation.

But is that a suitable goal? It is true that Ronald Reagan was elected espousing a certain approach to government and economics that he clearly has not been able to precisely carry out. But we do not really know if he was elected because of those proposed programs or because the alternative offered was less attractive to the public. In my opinion the election process is more complicated than is revealed by a surface glance and we really do not know, based merely on who won, if there is a mandate for a particular policy, or what. The voters also choose a Congress, quite often following a different path in that selection. Do they do so knowingly?

Our constitutional framers feared the tyranny of a suddenly formed majority reacting to some problem like the current economic malaise and making drastic changes that undermine the concepts of democracy and direction of the nation. They had come here from lands ruled by dominant leaders and systems that encouraged that dominance, and they were determined that the excesses that seemed to inevitably flow from such systems would be prevented here.

I suggest they were wise, indeed exceptionally so. They created a system whose greatest strength, not weakness, is its inhibitions against allowing a strong leader to arrange dramatic change. We have seen far more difficult times and survived them with our system and its curbs. Our wisest course today, I believe, is to continue to respect those limitations for they ensure liberty more than they inhibit fleetingly popular changes that risk an excess of power in mortal hands.

Thomas Jefferson advocated a living constitution, one that could change to meet the demands of the new times and he was correct. Ours has changed through mostly worthy amendments hammered out after serious and lengthy introspection, and when supported by large majorities as in the Roosevelt era. But now the cries are for a fundamental change, one that would disturb the tripartite equality of the three branches of our government and would allow an ascendant presidency. Resist!! The temptation may be great after a performance like the last lame duck affair. But the Madisonian concept that "ambition should thwart ambition" has served us well and should remain.

7/5/81# Say It Isn't So!

One of the oft repeated themes of this column is that communism, and in particular the Soviet brand of the genre, is a proven failure. This reality has received a lot of exposure lately through remarks by Reagan and Haig, the excellent CBS Special on the defense of the USA, and most eloquently in a marvelous book by the noted French author, Raymond Aron, *In Defense of Decadent Europe.* But as this fact becomes more widely accepted, a very important question is raised, why have so many nations turned to communism, so many young men and women fought so gallantly for it and with generally more success than their foes, and why, even in the heartland of the West, do significant numbers of people still vote for communist candidates or argue and debate over its ability to create a better world?

I have long pondered this question and have concluded sadly that to a considerable degree, *we have made it so!* The "we" I refer to is the USA and our predecessors as the dominant powers in Western civilization.

The primary causative factors have been two-fold. First, in evaluating the world the West has too often thought of it as like us, when in reality only the US, Canada, and some in Western Europe are generally alike; most of the rest of the world is characterized by far lower living standards, by incredibly greater gaps in wealth between the ultra-rich few and poverty ridden millions, by governments that represent the wealthy and/or the middle class only, and by leadership classes that dance to the tune of foreign powers and were placed in charge and propped up by those colonialists.

The second factor is the paranoia of the West toward communism. We are so obsessed with preventing its spread that we consistently side with the forces, any forces, that oppose change for we know that change means instability and the increased risk of a communist takeover.

The first factor blinded us to the desperate need for change in the rest of the world and we in the US, in particular, unconsciously encouraged that change by our rhetoric about democracy and equal opportunity and by showing the world, through the achievements of our society, how free, self-governed and affluent people could be. The masses were impressed and stirred to action only to find that their attempts to create change were fiercely resisted by the entrenched leaders and force was needed—as it was in another place in 1776. But the resultant fight terrified us because of its possible opening to the left so we sent arms, advisors, and dollars to the status-quo adherents to help them put down the revolts. The revolutionaries, our natural allies, were thus forced to turn to others for matching aid, and that move heightened our suspicion that they were Red and made us even more afraid of change.

But in spite of us, often the revolutionaries have won, and it was at this victory stage that we most contributed to their slide into the communist camp. Angry that they had defeated those we supported and even more sure that they were to become foes, we were ambivalent to the new rulers—not quite ready to write them off, yet; unwilling to offer more then minimal assistance for fear we would be helping future enemies.

Such an attitude is understandable, indeed predictable, but oh so self-defeating. Put yourselves for the moment in the shoes of a Nassar, Sandinista, or Castro who has just taken over a land that for years had been dominated by dictators and run primarily at the whim of foreign powers. Most of the wealth has been spirited out of the country, the land and industry is either owned by foreigners or by family members of the former "Numero Uno," and you have promised the people a change—land, jobs, more material goods, more say and opportunity—and you want to deliver.

But how? The country is broke, most of the educated people have fled, and the richest nations in the world are saying to you, "We won't help because you look to us a lot like a communist."

There is no successful model for quick nation building in the capitalist mold except by massive infusion of aid dollars, trained advisors, and private investment as took place in Taiwan, Japan, and South Korea. There are, however, numerous communist examples where the living standard of the poorest has risen, albeit at great cost in freedom and human dignity. So that is the court of last resort, but to whom among those poor millions do you give the land or a particular factory? There is no way to decide on a capitalist approach equitably, yet the people's demands for action are rising. You must do something or they will conclude you are no better than your predecessor. So you seize the land and the businesses and tell the masses that now THEY OWN THEM ALL. It seems the only way out, yet the very act slams shut what little crack remained in the door to the West. "You see," the cries ring out—"we told you, another Red devil has been aborne."

And so they become, for now only the Reds will assist them and that assistance leads to the inevitable bureaucracy a la the USSR, the corruption, the inefficiency and the drudgery. The once courageous revolutionaries are slowly coopted or they sense the failing and flee. Those that remain become members of the apparat: paranoid, disillusioned, and determined to stay in power at all costs and deny their failure. It's not hard to see where this process is most vulnerable to disruption and correction, but there is no sign of such a reckoning in the alleged Reagan foreign policy. Have we truly been resisting the Reds or helping them?

Words And Deeds
Should Match

9/18/83

The words and deeds of the Reagan administration certainly look like a mismatch as regards the tragic event of Flight 007. If indeed the Soviets are the vile evil force described by the president, as he claims the incident illustrates, then his actions seem totally inadequate. While if the incident were as mild as the action taken, it hardly seems to merit all of the rhetoric.

It seems to me that as a direct result of his imprisoning philosophy about communism the president is managing this incident poorly. He simply cannot bring himself to pass up this chance to beat the Reds over the head, yet, he also realizes our allies would never support his using the incident to break off relations with the Soviets, the only action that befits his language about them and the tragedy.

All in all it seems to me the US case has been weakened by our overplaying it. First the tapes were played for all the world to hear and the case was made that any US involvement was an absurdity. But then we had to backtrack and admit that a US spy plane was in the vicinity and that we routinely fly along Soviet airspace collecting intelligence. Then we claimed there was absolutely no indication that the Soviets fired warning shots before they destroyed the airliner, yet, belatedly we admitted that there was a possibility the pilot fired cannon shells before he launched the killer missiles.

Finally the president, ill advisedly I thought, stated that the Soviets were "stonewalling" about the issue. A word that reminds Americans and the world that an American president caught red handed with his hands in the cookie jar also stonewalled (lied) during the Watergate affair, an illustration that not just Soviet governments are reluctant to admit culpability.

All of this weakened a very good case by making it appear that Reagan wanted to make propaganda hay out of an incident that should have been apolitical. He made that even more obvious when in his speech to the nation he tried to tie the incident into support for his arms build-up programs.

Flight 007 is a sorrowful event that needs to be addressed on its own merits. The Soviets almost routinely use civilian airliners on spy missions and the West is not totally innocent of such use as well. Thus it seems to me if we care about preventing such future tragedies we need to find ways to diminish the mutual paranoia that causes us both to think we need to use any means to spy, and to protect our secrets at any cost in desperate fear of the other gaining some inexplicable advantage. Obviously not an easy problem to cure but certainly not correctable by rhetoric that perpetuates the competition and exacerbates the paranoia. I believe Reagan is wrong when he says the incident proves the US must spend more on defense. The airliner did not fall because of the recently gained Soviet military parity with the US. The Soviets committed equally despicable acts such as the invasion of Hungary, the Berlin Wall, the shoot down of US military aircraft near Soviet airspace, and the foothold in Cuba, when the US was the acknowledged world's supreme military power. The MX, Cruise or Pershing missiles, the B-1, M-1 tank, or any others of the planned Reagan build-up would not have saved one of 007's passengers. In fact their additions and matching Soviet add-ons will only make such incidents more likely. And such deplorable incidents have not been limited to the Soviets or other communists—just ask a Nicaraguan mother whose son has been killed by the "for hire contras" sent to that land by US dollars.

I am not a seeker of unilateral disarmament. If I thought the US was weaker than the USSR I would not oppose the new weaponry. But we are not, the US and its allies possess the most formidable military force in the world today, yet, we were unable to secure the passengers of 007 or bring liberty to Poland. Our greatest weaknesses today are not weapons but our weakened economy from the huge budget deficit and our fascination with power and the obsolete belief that in the nuclear age superiority can still be purchased and determined by bean counts of weapons. 'Tis a shame.

Power And Security Do Not Always Compute

6/17/84

Both the president and some of his supporters are fond of stating that he has now strengthened the USA and that we now live in a safer world. Would that it were so, but 'tis not. I usually ask such advocates to be more precise and tell me what it is that Reagan has done to make the US strong again, and what actions the Soviets have allegedly backed away from as a result of our new found strength. They reply the MX, the B-1, the Euromissiles. Then I remind them that we have not deployed any MX's or B-1's, and only a few of the Pershings and cruise missiles, hardly enough to alter any military balance in Europe. Next I press my case about what Soviet actions we have averted by this meager new strength, and of course, my compatriots are bumfozzled. Obviously one can never assert with any certainty what they have prevented, for since it was prevented, it did not occur. Thus they usually can only suggest to me that certain things that happened before Reagan would not have happened after his succession. Afghanistan is the favorite of such asserters.

I scoff at the reply. Can anyone really believe that even a whole force of B-1's, MX's, Tridents, Euromissiles, or New Jersey type battleships would have altered the Soviet decision to invade Afghanistan? In order to so postulate, it seems to me, one must make the case that such weapon systems are, at least remotely, probable for use against an aggressor who might invade Afghanistan. But is that not preposterous? Would, under any circumstances anyone can reasonably conceive, the US have considered going to nuclear or any other kind of war with the USSR over Afghanistan?

The US was not willing to go to war with the USSR over Hungary when we had an almost absolute and indisputable nuclear advantage, something we could not claim at the time of the Afghan invasion, nor now, at the beginning of the Reagan build up. Nor were we willing to do so when the Soviets invaded Czechoslovakia in 1968, when we were also stronger vis-a-vis the Soviets than now. The truth is that the use of US super weaponry against the Russians as a result of their invasion of Afghanistan was and is an absurdity. Thus the acquisition of such weaponry obviously would not have deterred the invasion.

People only wishfully make such remarks. Remembering what they thought were the good old days, and which never really existed, when the US superpower dominated world affairs and no actor moved without considering our reaction. But that is all a myth and one that causes this nation no end of trouble. Too many American officials, as well as just enthusiastic supporters of our ethnocentrism, think only in such "zero-sum" terms, assuming that before any other nation acts it must first consider what the US will do about it, and that if they act unwisely in our view it is somehow a US failing. But it's not so!!

Most nations, and especially the other major powers, have many interests which they consider outside the realm of US and most other nations' influence. In dealing with such interests they act exclusively as they deem necessary by the circumstances. If tempted to do something audacious, they may pause and consider whether the action might lead to a nuclear confrontation but once they are assured that is not likely, as it clearly was not in Afghanistan, a country long in their sphere not ours, they will act regardless of the wishes of the USA.

The USSR feels, with considerable evidence in support of that belief, that Eastern Europe and most of their neighbors fall in this category. Since Reagan has been in office the Soviets have shattered the dreams of Solidarity in Poland, and Reagan was no more able to prevent it or alter the outcome than Carter, in fact he was forced to do what Carter did, minor economic sanctions. It can also be noted that in spite of our alleged increased strength, the Soviets remain in Afghanistan, shot down the Korean airliner, strongly rearmed and supported the Syrians during our ill fated effort to restructure Lebanon, and they continue, in spite of the veiled threats of our administration, to support, we do not know how much, the revolutions in Central America and elsewhere.

It's high time Americans and their leaders recognize that merely adding armament to an already excessive arsenal will do little to affect the actions of our rivals. When we learn that, we may be able to reopen the dialog and return to the improved atmosphere and potential of the mid-'70s, when we really did live in a safer world and when the weaponry of Armageddon was starting to slow rather than accelerating. Part of the process of getting there will require us to quit thinking that we have the power to shape all the world's events and that any action the Soviets take must be a result of our shortcomings, a plus for them and a minus for us, and vice versa. It's a distortion of reality.

3/11/84
Agree With Reagan

I do not know if I can do it. My fingers have become stiff and my palms wet and clammy, but I will do it. I will write the words, I AGREE WITH RONALD REAGAN. I agree that a tax cut is not the answer to the budget deficit, and that eliminating or delaying the tax indexing for inflation is a cowardly way for government to take from its people. But I also agree with the Reagan critics and his own economic advisor, Feldstein, and the fed chairman Volker, that something should be done about the deficits now. We cannot afford to wait, as our president proposes, until after the election. To me that is unacceptable leadership.

How did we get into this mess after we elected a president, just three years ago, who promised to cut government spending, reduce taxes, balance the budget in his first term, and create an economic recovery without resorting to high unemployment? Not incidentally he also promised to increase military spending and that has turned out to be the rub. For in a nutshell Reagan has increased military spending more than he has cut other spending, while simultaneously his tax cut program failed to rejuvenate the economy either as fast or as effectively as he promised.

But Reagan obviously does not deserve all the blame. Another major contributor to the rapidly increasing cost of government has been the skyrocketing bill on entitlement programs. These programs, which consist primarily of Social Security, Medicare, and federal type retirement benefits, will grow more than 45 billion in constant dollars between 1983 and 1988. But the major reason for their growth can be traced back to 1973 when another administration, but note also Republican, in consort with a Democratic Congress, decided to index Social Security and retirement benefits to the cost of living indicator. As a result the costs of these programs have grown faster than any category other than defense.

One might argue that Reagan's defense spending will at least spur the economy and decrease unemployment, but the facts are it will do so less than tax support of other programs. According to Les Aspin defense spending of a billion dollars employs only about 35,000, while a similar amount for public service employs 132,000. Additionally defense spending has a lower multiplier effect.

Deficits can be reduced in two ways, bringing in more revenue via some form of taxation or taxation collection improvement, or by decreasing spending. It is interesting to note that past tax increases have not reduced the deficit. For example even Mr. "I Hate Taxes Reagan" endorsed a 100 billion dollar tax increase in his second year but the deficit grew.

This tells me a simple truism. If you give money to government, any government, from any party or segment of the political spectrum, they will spend it. They may spend it differently but they will spend it. Thus the only safe way to reduce the deficit is through reduced government spending.

Now the obvious place to do this is in the area where Reagan has refused to cut and in fact has provided for the greatest peace time growth in our history. In 1980 the defense budget was 136 billion dollars. For 1985 Reagan is seeking more than 300 billion, over a 100 percent increase in a mere five years. From 1983 to 1988 Reagan wants 154.5 billion constant dollars more for defense, a growth that is greater than the total cost in 1980 and will represent 63 percent of the total budget growth for the period. This is absurd for the threat has not changed. A no growth defense budget would thus save us three-quarters of the projected deficit by itself. Note that a no growth defense budget is not an anti-defense budget, it would still leave the US spending almost 70 billion 1980 constant dollars more on defense in 1985 than in 1980, and leave us with likely the highest defense budget in the world.

By combining a no growth defense budget and a change in the indexing system for Social Security and retirements to one point below the average wage or the cost of living, whichever is lower, we could significantly reduce or possibly even eliminate the deficit in the next four years. If we added to that a no growth limit to the rest of the federal budget (adjusted for inflation), we could surely win the battle of the deficit without increasing taxes and taking more away from you and me. Its your government, speak out and make it happen.

CHAPTER EIGHT

Ethics

The citizens and leadership of the USA are caught up in a difficult ideological struggle between those who want this nation to be different, special, and indeed loyal to its initial ideas, and those who want it to be practical, pragmatic, realistic and follow the past precedents of previous nations who found themselves at the top of the pile, king of the hill. We have been treated to contrasting examples of the two approaches very recently via the first two years of the Carter administration, the Reagan administration, and the sort of blend offered to us by the Nixon/Kissinger years. The answer as to which of these alternative guidelines the American public really prefers deep down in the deepest recesses of their hearts and minds is currently very much in doubt.

I conclude that it is in doubt because the American public initially responded favorably to Carter's promise of putting human rights and ethical considerations back at the forefront of American policy, only to turn against that man when his opponent in the next election made it appear that his policies had weakened us vis-a-vis the USSR, and made it possible for the Soviets to take advantage of our wimpishness internationally. The media and academe may bemoan Reagan's near return to cold war confrontation and an emphasis on the use of the military element of power as the lead-off hitter in each round of international hardball, but the general public, as indicated by their reactions to Grenada and Lebanon, does not seem to concur.

However, that same public when asked specific questions about Reagan policies, seem more often than not to disagree with the president and lean toward the kind of policies Carter emphasized when he was running for election and conducting affairs in his first two years of office.

This willy nilly wavering of the public reflects a conscience and a preference for our idealism, but only if it does not run any risks. Like most we seem to be willing to be ethical until it might hurt our pocketbook or our security, then we prefer realistic toughness. Thus, the public seems to say to Reagan, "Don't invade Nicaragua, but sure give 'em hell short of that," although if asked specifically if we should support the Contras, those who know who they are, by a strong majority, prefer not to.

This very issue is dealt with in a classic book with a new edition about to hit the stands as I write these words. The book is by Hans Morgenthau and the latest revision was completed recently by his long time associate and student Kenneth Thompson. The book is entitled, POLITICS AMONG NATIONS: THE STRUGGLE FOR POWER AND PEACE and it has probably been the most influential book on American foreign policy since the end of WW II. It was Morgenthau who lured the USA away from its earlier belief that power politics was something to abjure, and that the game was better played, if at all, with morality and ethical considerations as the major element of it. Morgenthau simply posited that between nations there were no ethics or morality, only interests, and those interests were decided by the accumulation of power and the use, or perceived intended use, of it. When Morgenthau first proffered this concept for America's guideline it was opportune because this nation, although for a long time had claimed to want to stay out of the forefront of global activity, suddenly found itself the most powerful nation on the globe, both militarily and economically and the sole possessor of the most awesome weapon in history. Also we had just learned that our idealistic views about global politics were not shared universally by the other survivors of WW II, and that our legalistic, sovereignty diminishing faith in international organizations like the newly emerging UN was running into the stark reality that most nations, including our own, really were not ready to let an international body tell them what to do. Thus, Morgenthau's views prevailed simply because they were easier and less risky, and did more to ensure America's newly earned favored status in the world than any alternative.

Early on there were critics. My favorite was the erudite William Fullbright, often called "Halfbright" by his detractors. But his book, THE ARROGANCE OF POWER, suggested that America had fallen under the allure of power and turned away from its idealism at the very moment when the world needed the American ideals more than ever before, and certainly more than a Pax Americana based on power and support for the status quo that made most of the world a far less attractive place to raise children than in the USA.

It is hard to say when the Morgenthau theories began to win out. It may have been the Berlin Blockade, the rape of Eastern Europe, the Korean War, the failure of the UN to be the substitute for force and institution that could provide peaceful resolution of nation state differences, our replacement of the Shah after the Iranians tossed him out, the successful overthrow of the Arbenz government in Guatemala, or any number of successful, more pragmatic, if unidealistic actions we managed to engineer. But clearly the power relationship ideas did supercede the idealism, and all around this globe a beautiful dream that had captured the imaginations of millions began to fade, became jaded, tarnished and finally, today, I suggest, cracked and barely eligible for refurbishing. The dream our rhetoric created around the globe after WW II and at the beginning of the late '40s, a vision of a world of rising expectations for the poor and middle class as well as the rich, a concept of political enfranchisement of the masses and a chance for upward mobility both in wealth and political power for almost anyone willing to give it a try, suddenly began to look like a lie as reality crept in and our support of the dream diminished. We had suggested that equality, liberty, upward mobility, influence, and basic human rights like a vote, a job, choices, education, etc. soon could be available to all without wars or revolutions or difficult and murderous struggle. But when that did not happen and people began to demand it and try to seize it, instead of siding with those that we should have considered our co-conspiratorial dreamers, the USA, more often than not, either sided with those with the power or opted out; thus in actuality siding with those who held the reins of power. This was true whether one looks at Central America or in Eastern Europe. America seemed to sit on the sidelines or join with the oppressors against the powerless.

Stanley Hoffmann of Harvard, a man I respect a great deal, suggests that America adapted Morgenthau to its own special creed. That although we were willing to use his theoretical support for playing the power game, we misunderstood or ignored the underpinnings of his theory, and thus undermined the effectiveness of our efforts at power realities. Hoffmann notes that Morgenthau insisted that in playing the realist game of power one had to recognize the limitations of power and very carefully establish a nation's priority of interests. This, Hoffman suggests we failed to do. We falsely assumed that we had unlimited power, over calculated our interests and tried to do more than the realism suggested by Morgenthau would allow. This explains Morgenthau's disillusion with US involvement in Vietnam and other US international shortfalls.

But I disagree. I believe it was the Morgenthau concept that provided us the rationale to storm into Vietnam, into an economic and military hegemony over Central America, and into a blind spending spree on military weaponry and the desire to use it and maintain some mythical nuclear superiority over all others. For Morgenthau posits that Machiavelli was correct, that there is no room for ethics and morality in world affairs, that sovereign nations are an authority unto themselves, and that the only way to resolve issues between them, and only then temporarily, is by power arrangements. He does not abjure diplomacy, in fact he approves of it over the raw use of power, but he argues that diplomacy must be backed by power calculations if it is to be effective.

In other words although I am an admirer of Morgenthau for the skill of his pin and the eruditeness of his persuasion, for his belief that political scientists are very wrong when they ascribe to the model theories of predicting world and interstate events, I also believe that his overall theme was that the USA should turn away from its idealism as it grows up in the world and become more cynical and pragmatic. That it should indeed adopt and adapt for these times the approach to international relations of Sparta and Athens, Rome and Great Britain, China and Egypt in their historical precedents, and act as a great power should with power in the forefront and our own individual interests as our guides.

And I believe that is wrong, antithetical to our future, and counter to our heritage. For if we follow the Morgenthau dictate, and I truly believe we have for most of the post-WW II era, merely adding the rhetoric of ethics and morality to it for cover, as most of our predecessor great powers also did, we will doom this globe to a fate far worse than it deserves if led by wise people.

If we pursue this so called pragmatism we will not change the world a whit, and it seems to me the measure of a nation's success should not be how long it survives at the top of the heap, but what it does to the heap while it is atop. If it does not leave that heap better than it finds it then that nation has missed an opportunity. One of the articles in this chapter refers to a newspaper column of Sydney Harris, my favorite of all of the many excellent columns Mr.

Harris has blessed us with. In it Harris, I believe, correctly and briefly describes the history of nation-states and the power concept and tells us that once on top no nation or group of peoples has essentially acted better to those beneath them than any other. Think about that statement! If true it tells us that the power concept is the only reasonable one to follow. But it also tells us that as people we have not gotten any better after centuries of thought and wisdom. We are no smarter than Erasmus, and just as ineffective at changing the animal-like or jungle-like realities of our existence.

I have concluded that Harris is correct in describing history, but I hope that he is wrong in describing the future. I believe that due to the existence of the nuclear weapon, the closer interrelation of the nation states via the communication revolution, and the rare opportunity of a top dog, a US that already has enough space, materials, and wealth to be virtually free of want if it acts wisely, has the unique opportunity to break that chain of history and act more wisely than our predecessors. But not if, as Morgenthau suggests and our leaders tend to accept, we continue to conclude we can only be respected for our power and stay on top by the periodic demonstration of it. Not if staying on top alone is considered the ultimate achievement.

This chapter questions the ethics of our national psyche as it applies to international actions. In the final chapter on solutions I will suggest an approach that I believe does not ignore power realities, but that allows our nation to act differently, and reasonably securely, to disavow the rules of the power game and nudge the nation state arena to a higher plane of international, rather than national, survival and prosperity. Read them with a cynical questioning attitude and reject them if naive or dangerous, but consider them and ideas like them of your own if they touch a chord and suggest a hope for a better reality.

3/18/84   # The Gall Of The Pious

The Reagan administration serves as an example of why a strict separation between church and state must be maintained if a government is to be judged objectively. For this administration, by seeking support for a constitutional amendment allowing prayer in public schools, attempts to clothe itself in Christian garb while simultaneously acting in a most immoral nature. By simultaneously seeking in the Senate the prayer amendment and a 21 million dollar bill to support mercenaries now carrying out attacks against the people of Nicaragua, the administration shows us how religious piety can be used to cover up blatant transgressions.

The crowning gall of this allegedly pious leadership is that when their bill for support of the "contras" was defeated in committee, they unscrupulously attempted to attach it to humane bills for feeding the victims of a drought in

Africa and emergency energy assistance for American poor. They did this in order to force more honest and sincere people to support murder if they wish to help the needy. So far their tactics have failed, but just the attempt tells us something about their total lack of understanding of morality.

This latest act of the administration is an example of what was termed by insiders during the Nixon days as "playing hardball." The term attempts to distinguish between ordinary folk, who are forced to play "softball," since they do not possess the requisite skills or unmitigated determination to be big leaguers, and those that are truly capable and when necessary, uninhibited by ethics.

Nixon and his team excelled at hardball and they were clearly proud of it, and thought that the American people shared in their admiration of a winning style. That belief led to Watergate, the coverup, and the attitude exemplified by the Nixon statement that if the president authorized it, it could not be against the law.

The sad part about all of this is that in some ways Nixon was correct. Americans do like to win and far too many of us think the idea of being a good loser is just for losers, which we are not. Winning thus becomes a goal that merits actions we would normally abjure. So in spite of the post Watergate revelations about America being involved in dirty tricks all over the globe, and the subsequent attempts by Congress and Gerald Ford to prevent such actions in the future, "America is back." Back at playing hardball around the globe by supporting men like Mobutu and paying for, training, and arming assassins who attack villages, burn bridges, and even bomb targets as volatile as oil storage facilities. And such decisions are not made by our Congress with its diversity and nearness to the American voter, but by one man and the people he appoints. Such decisions are seemingly made easier by a self-ordained religious zeal authorizing any action against evil leftists.

Why does the administration not have the guts to ask Congress for a declaration of war against Nicaragua and then, if authorized, launch an open attack on the Sandinista regime? Because it knows from past experience that Congress, more closely attuned by its size and proximity to the people it represents, would refuse. So they prefer to work in semi-darkness, a place where evil usually flourishes. And like all their predecessors who have dirtied their hands thusly, these hardball players too are becoming sullied by their black ventures. Desperate, when stymied by men more true to our nation's heritage, they now stoop to attaching riders of darkness to bills of light, and to trying to force more honest men and women to authorize the use of fire and pestilence before they can help the poor and hungry.

When such men support public prayer we would well be warned that someday they might translate public support for that seemingly innocent prayer amendment as equal support for their violent means. Intelligent men and women, whether religious or not, should deny any opportunity to connect their religious beliefs with or to their governors. I oppose the prayer amendment while I heartily endorse prayer, and I oppose the war on Nicaragua and the perfidy of connecting it with other, and more honestly humanitarian, goals.

 # A Hostage's Xmas Message

Sydney Harris, my second favorite columnist, once wrote that the world is a seesaw with some group always on the way up while another is coming down. But the really sad part, he argued, was that each succeeding group acted just as badly to those that were down as they had been treated when they were there. Having success seems to make us hate those who also want it. Harris has probably correctly described history, but I hope not the future. I have always felt that someone will come along and reverse that trend, and naturally I have hoped it will be my own nation. Something I read recently adds to that expectation.

It was a letter to the editor of NEWSWEEK written by a former hostage in Teheran, John Limbert, and in it he criticized the magazine for accusing a Shiite Muslim group of being the "mad bombers" that hit our Marines in Lebanon. In my opinion he properly accused Newsweek of printing allegations without sufficient proof, and in so doing perpetuated "misinformation" about the 50 million or so Shiites, caricaturing them as a "gang of crazed and fanatical murderers."

Mr. Limbert went on to note that Shiites are not unusual in their willingness to become martyrs for a cause, even Christians have done so and have been honored for the act. Obviously, John Limbert is a man of unusual perspicacity and forgiveness. Traits our nation could benefit by adopting.

The Christmas season seems to be a perfect time to raise the question, why are we so eager to hate and disapprove of others, especially those who differ from us in culture, language, belief and appearance? Why do so many Americans, like our president, assume and popularize the idea that communists are evil? Why did so many think that the only way to react to the hostage seizure in 1979 was through violence? "We should nuke them and turn Teheran into a radar oven," I was often told both then and now. This easy hatred of those who differ from us makes it easy to support the spending of our coin more on guns than food, education and health, even though clearly the world has more guns than butter, more suffering than prosperity, and the suffering is a greater enemy to us than the guns.

It is prevalent everywhere. Argentinians revile the British; Soviets call us capitalist war mongers, and we call them the focus of evil; the Ayatollah Khomeini refers to the USA as the Great Satan; Turkish children grow up hoping for the opportunity to kill Greeks; and North and South Koreans, people almost from the same womb, brutalize and defile their brother/sisterhood.

It seems clear to me that Christ's message was different than what we are practicing. Unlike his predecessors, he did not emphasize "an eye for an eye," that God was behind the wars "His" people might wage against the infidels, and that "He" would aid us with pestilence and fury against the ungodly or wrong Godly foe. Instead he stressed love, even of thine enemy, forgiveness, understanding and caring. But the self-proclaimed prophets who lead many of us seem confused and are instead making it easier for us to follow the Old Book rather than the New and thus, make hatred and violence seem reasonable.

Congratulations, John, you obviously learned much from your days in Iran. Your broadening experience living among those who by a twist of fate changed from friends to foes has stood you in good stead. Your wise counsel in defending them reminds me of another John who walked among us many moons ago and courageously interpreted the teachings of Jesus in a way that was not too popular. But neither his efforts nor yours seem to be reaching the majority of us, the people of this globe, or our leaders. The hatred and bitterness, the desire to be sure we are better, stronger, more like God than the others, seems too overpowering.

But it does not need to be. Mr. Limbert's message is an important one to examine this Christmas season. He shows that we can use our intellect to understand, to forgive, and seek a modus vivendi even with those who differ from us and our tenets. I like his message better than most that I hear this Cristmas season and hope that many of you will ponder it.

10-3-82

The Lessons Of Force

Lost in all of the hullabaloo over gun control versus constitutional rights to possess arms is a far more important issue. Why do Americans use guns to kill one another more frequently than other similar societies?

Our neighbors in Canada offer a striking case for comparison. In 1980 they killed only eight of their compatriots with handguns while in the USA more than 11,500 met the same fate. America's population is about ten times that of Canada but the murder rate is 1400 times as high. Are Canadians automatically less violent than we, more normal, or what?

Well, no reflections on my Canadian neighbors are intended, but I cannot accept such a seemingly obvious explanation. I feel more at home in or with Canadians than in or with any other foreigners. Both of our societies' majorities speak a similar tongue, we both live in a democracy, have states' rights, a mania for automobiles, and enjoy baseball, football, hockey, and many of the same TV shows, entertainers, and foods.

Of course Canadians have a national gun control law which makes the ominous little critters harder to come by, but Canadians, like us, love to hunt, are emerging from a frontier society which strongly emphasized individuality, independence, and the right and expectations that one would scrap for the golden ring. So our likenesses seem to far exceed our differences and deepen the mystery about the handgun murder difference.

Many attempt to explain the US high overall crime rate as well as our murder record by pointing to the capitalist system we follow and the alleged greed it supposedly inherently generates. But if that is the cause, why not the Canadians too? They are capitalistic.

Detroit and Windsor, Canada, exist almost side by side divided only by our nearly unpatrolled border, yet, while Motown is nearly overrun with

135

crime and has a shocking murder rate, Windsor is quiet and relatively speaking offers a near asylum-like safety.

So we return to the easy accessibility of some 55 million handguns in the USA. But to accept that explanation is to conclude that murders are seldom premeditated, and that Canadians would be just like us if the tool was as readily available to them. And I can not do that. I am too impressed by the argument of the gun lobbyists, with whom I disagree in their opposition to gun regulation and licensing, but whose logic that guns do not kill, people do, strikes me as amazingly convincing.

If their point has any validity then gun legislation becomes like chemotherapy treatment for cancer—a necessary resort to a painful treatment in order to save some of the victims, but hardly a preventative for the disease and its spread. And it is spreading—Americans killed more of one another with handguns during the peak seven years of the Vietnam war than the Viet Cong combat operations could match.

My attempts to find a solution for this violence syndrome in our land was given a rude shock when I discussed the issue with several foreign friends. They all agreed on what appears to be a simple explanation that springs from a common international view that Americans are an especially violent people.

They suggested to me that people follow the examples of their leaders, and that our leaders are most often quick to resort to guns as the solution to almost any frustration or problem. It matters little if the peasants revolt in Central America, the middle class demonstrates in Europe, or our embassy is seized or attacked; our nation's response, they suggested, is almost always to send guns, or add guns to our arsenal, send advisors to teach the locals how to use guns more effectively, send troops with guns, or show our resolve to meet this issue by an increase in defense spending and cutting back on human services.

A change in that leadership prescription to most problems may well prove more valuable in saving lives from handgun loss than any move to limit gun sales or possession.

EUREKA! A difference between us and the Canadians not restricted to ease of accessibility. The Canadians spend little on armaments, are far better known as peacekeepers than as fighters, negotiators and compromise suggestors rather than supporters of one faction against another. They seldom export arms and even less often sound the claxon of tough military bluster.

Could it be that those who judge us from afar have seen us more accurately? Has Reagan's cowboy rhetoric, his increased military spending when almost all else is under the knife, his solution of guns to El Salvador, Argentina, Zaire, his revival of the capitalists' weapon (the neutron bomb), et. al. perhaps even contributed to Hinckley's attack on him? And did his predecessors' false missile gap and the Vietnam War, the unleashed CIA's excesses all combine to set the wrong example for the less stable of our own society to emulate? If one becomes famous, honored, admired, and well off by dealing with problems of utmost importance through force and power than why should not the assassins of the Kennedy's, Wallace, Reagan, and the hundreds of John and Mary Doe's also become famous and resolve what they perceive to be major problems and/or frustrations through their flimsy handgun imitations of America's use of its military might?

There is little doubt that Americans shoot one another in horrendous numbers partly because the guns are there, but do they also resort to such acts because every day they receive a subliminal message from their honored heroes, that resort to force is a way to earn respect and success?

Mothers And Their Offspring

5/8/83

"ANY MAN WILL BE HAPPY WITH THEM BECAUSE THEY HAVE BEEN RAISED TO SUFFER." Those lines used to describe daughters in the brilliant new Garcia Marquez novella, *Chronicle of a Death Foretold,* have been haunting me ever since I first read them. Mother's Day, 1983 seems to be an appropriate time to share my thoughts. For the Nobel laureate was writing about mothers and wives and the rather depressing condition of humankind. The book weaves a story about the plight of Latin America, about how helplessly it has enslaved itself to custom, superstition, machismo, honor, fate, and distorted community standards of right and wrong. But I suggest the book tells us as much about the whole world as it does just about the Latino part of it. For although Latin America may serve as an excellent laboratory for observation of the evils of old mores, superstition, and closed minds, the people who reside there are really not much more guilty in holding back the reasonable progress of human thought than their brothers and sisters all over the globe. We may find even more shocking limitations on reason in far off places in Africa and Asia, but many ridiculous and ultimately evil roadblocks to human soaring exist even here and are accepted as normal.

The Marquez story tells what happens in a small Latin village when a young girl is returned to her family, after her wedding, rejected for not being a virgin. Her brothers are driven by the societal norm to demand the name of the violator and to send him to his appropriate death. She gives a name, it is never verified as being the right person, but is assumed to be accurate and the brothers, reluctantly but inevitably, commit a violent murder—anticipated and in varying degrees almost relished by the entire community and condoned even by the eventual trial. No one in this story comes out very pretty, and I believe that is the way Garcia means it to be—this is clearly not meant to be a celebration of humankind's glory. I believe Garcia is saying to us all—is it not time to wake up to all of our shortcomings? Is it not time to shed the superstitions, the superficiality, the false role pictures for brother, sister, man and woman; to cease glorifying jingoism, the act of murder and other prejudices by sometimes calling them justified, indeed required to right some perceived wrong committed by one who violated our view of what is right.

That description of the reality of life seems to me to offer a worthwhile issue of consideration on a national holiday honoring women for being mothers. We ask a lot of them and they have delivered even more for centuries. Yet men have made the rules that both they and the women have had to live by, even though many of those rules, often as earnestly clutched to their bosom by women as well as men, Marquez suggests, have crippled humankind from achieving its best.

Oh, those accepted practices have seldom prevented us from material successes but they have damaged our souls. Made too many people feel valueless and too many others pay an irrevocable price of social expulsion for minor transgressions.

If there is one thing that mothers honor it is their offspring. Yet mothers, perhaps more reluctantly than men but nonetheless too, have been persuaded by society to accept a standard of just killing. The ultimate symbol of motherhood, the Virgin Mary, in fact, has become a battlefield charm as soldier after soldier goes off to do battle assured by the virgin around his neck or painted on his rifle barrel that he is doing right.

You have been too quiet mothers, too acquiescent to the mores that men have weaved to cover their baser instincts with innocence. Women in much of the world are finally speaking out for their economic rights, but in so doing run a great risk that they will become only more like the men who created the fabric of the society we reside in. Perhaps this will be better than the more acquiescent role of women portrayed in the Marquez novella, but not by much, for if women just join the males in their rituals and superstitions, always thinking that the old way is the right way and never questioning it, while allowing themselves to be satisfied primarily with the material successes that have clearly occupied men's minds; then we will prove Marquez to be a flawless prophet of the death foretold. The death, if not OF humankind, than of its worth, and we will thus reign on this earth for a bit longer but only at a slightly higher order than beasts.

But if on this Mothers Day all mothers vowed to step forward and demand that this earth be run as if all people on it were their children, and indeed they are, and that murder, pride, honor, self-righteousness, wars, conceit and all of that which we use to conclude that one group is better than another, from which one group determines it has to destroy or frighten or bluster against another, are to be abolished—well it would indeed be a day to honor. How about it mothers????

4/7/85

The Easter Message

Today we celebrate an event that occurred almost two thousand years ago. A happening that has touched the lives of more people than any other incident in history. We will celebrate it with prayer and songs, our Sunday best clothes, an abundance of good food, and other events centered around our prosperity. Many a minister will remind us of the "true" message in Christ's life and resurrection, and we will return home to our comforts and plenty. But the real lesson I see on this Easter Day, I am sorry to say, is how little we have really changed as a result of that first Easter Sunday.

We are told Christ died to redeem us from our sins and to teach us about a new life of love, sacrifice, concern for our neighbors, compassion, forgiveness and all of those wonders that we preach about often, yet pay attention to seldom. The world he visited temporarily was a sinful world full of personal ambitions and endeavors for wealth, national rivalries, wars, and enormous gaps between rich and poor. The world of today is not noticeably different.

It is not often emphasized these days but the truth seems to be that Jesus spent most of his time not with folk like most Americans but with the poor and downtrodden, the barely educated, those disenfranchised from power, wealth and success. It is they whom he taught, succored, and for whom he sought a better fate.

And how are we, his followers of the 20th century, following his lead? Well, look around at the splendor of the USA; its affluence and achievements. It is so amazing one can almost overlook the pockets of poverty and despair that have worsened for the first time in decades these last four years. But if one projects a wider view, indeed a global one, for Jesus told us we were our brother's keeper, the picture worsens. For on far too much of this sphere we call Earth there is greater poverty and sickness, greater ignorance and deprivation, both material and political, than existed in Christ's day.

Charles Maynes, the editor of FOREIGN POLICY and a former Assistant Secretary of State, recently was quoted in the WASHINGTON POST thusly: "The world is on the verge of a human catastrophe and political disaster that this country seems determined to ignore.... Whole countries have seen their hopes for the future disappear...." Maynes went on to note that the US was now competing with Italy for the dubious honor of being the least generous of all western countries in the provision of aid to the needy, while using its influence to reduce the World Bank's assistance to the poorest of the poor.

The WORLD DEVELOPMENT FORUM, a publication of the Hunger Project, noted in February 1985 that Costa Rica's relative success in comparison to the rest of the developing world can be traced directly to the fact that it was isolated and poor and without a cheap labor force to be exploited during the 300 years of colonial domination. Most of those colonialists were Christian and most of the nations they dominated are still suffering from that imperialism and its economic aftermath of cash crops, landless majorities, and dictators beholden to a small elite who still cater to the whims of the rich from afar.

In the meantime in 1985 for the first time world military expenditure will top a trillion dollars, led by a Christian fundamentalist USA and an atheist USSR. And the USA and the USSR both all but ignore the pleas from the Third World for assistance via a steady and long term financial and political commitment to their needs. Our combined military budgets approach trillions while our aid to the truly needy rarely reaches a thousandth of that figure. We have obviously learned somewhere, but not from that cross, that it is more important to be able to kill than feed, clothe, or medicate.

If that so called Prince who died on Calvary were among us now could he let us feel smug and complacent on this Easter day? I think not. Happy Easter!

1985   College Athletics Need A Fix

American college athletics seems to be caught up in a recurring nightmare of violations, half-hearted remedies and repeated scandals. This year we have the Tulane point shaving incident along with the usual coaches resigning because they say the only way to win is to cheat. There are drugs, phony admissions and eligibility, payoffs and about to graduate athletic standouts who seem unable to read a book. All of this to be resolved this month, of course, by another meeting of college presidents. Don't hold your breath!

In my opinion what is wrong with American college sport can not be remedied by rule changes and the kind of symptom treatment being recommended. What is needed is a major overhaul that strikes at the very root of our system.

America is the only major sport nation whose amateur athletic programs are rooted in education. We, almost alone in the world, seem to believe that academics and athletic prowess go hand in hand and that misbelief may well be the major cause of our continuing hypocrisy. Our insistence on this illogical connection discriminates against many of our potential athletes and contributes mightily to the athletic excesses that tarnish our colleges, universities and even our high schools. Blemishes that most other nations avoid simply because they develop their athletes via sport clubs rather than educational institutions.

How does the system discriminate against athletes? First it denies many of them entry into our college system, even though the colleges, almost by accident, have become the primary development ground for future professional and world class amateur athletics. Many are denied this development opportunity because their high school grade point is too low. Since there is no connection between a college degree and athletic field performance, this is clearly discriminatory.

Of course if the athlete has demonstrated superb promise on the field some way is usually found to sneak him by the rules, but that cure is worse than the discriminatory situation that creates it so a new potential scandal is born.

The system even discriminates against those athletes who are legitimately eligible and who strive to garner an education along with the development of their athletic skills. For they are held to a different standard than most other students. They are generally required to major in some subject quite remote from the real reason they are in school, their sport. The excuse for this being that the likelihood of their making it in professional sport is so remote. But other students are not so directed and limited—they can major in subjects with little likelihood of future employment, and indeed most college graduates do not end up working in the field they studied in school.

Additionally athletes, unlike most students, must carry a minimum number of credits and they alone must make certifiable degree progress, whatever that means. All this while working out 3-5 hours a day, most often both in season and out. They are closer to indentured slaves of the institution than normal students. So what to do about it? The answer seems simple. Switch to the club system and disconnect academic and athletic performance. But what happens to that wonderful and revenue generating US college athletic tradition? Well, if we play our cards intelligently, nothing. We simply create sport clubs on college campuses. The clubs, probably run by alumni, pay a minimal fee to the school for the use of its logo and facilities and continue the athletic programs, but with important new freedoms.

They can recruit any athlete they wish regardless of his/her academic standing in high school. Probably there would be an age limit or years of eligibility limit, but even that may not be necessary. They can pay the athlete what they wish depending on the national amateur rules which allow compensation for missed work while practicing and performing. The only requirement would be that they must also give the athlete one year's tuition, room and board to the club's host school WHENEVER HE WISHES TO USE IT AND IF HE MEETS THE SCHOOL'S ELIGIBILITY REQUIRE-MENTS at that time. The athlete then has a choice. The athlete can go to school while playing for the club, if eligible, but his play will not be determined by his grades. Or he can use the scholarship later after his playing days are over and there will clearly be more time available to concentrate on the studies.

This change would remove most of the reasons for cheating and misguided actions by our college and university athletic participants, retain the traditions of the past, and meet the new reality that college sport has become, a business and a development program, not a mere game between regular college students. Any other suggested solution retains the illogical connection between sport and academe and thus will eventually fail.

CHAPTER NINE

Solutions

One of my favorite pet peeves is the critic who is always willing to tell us how badly we have managed things but who offers no alternatives. My columns therefore often featured alternate policies and discussions of their pros and cons. Many readers tell me they appreciate that fact above all the other facets of the series.

Still it is imperative to point out that in the field of international affairs seldom is any issue totally resolved. One merely has to glance at history to confirm that discouraging reality. Almost everyone agrees that WW I led to WW II, and who can deny that WW II has set the stage for much of the world's problems of today: Israel's creation to compensate for Hitler's atrocities against Jews and the resultant instability in the Middle East; the rise of NATO and the Warsaw Pact out of the occupation zones of the US and the USSR; etc.

All of the above is said to suggest that the reader must not expect too much from this chapter. Solutions are merely steps that begin to deal with some of the world's major problems, but in the international milieu each step of the solution is inevitably going to create new, although related, problems, and so the world goes on, dampening down one crisis in preparation for the next, bandaging over and jerry rigging while the next unexpected dimension springs forth.

The major focus of my solutions is to bring the reality and the rhetoric of nations' actions closer together. I have the crazy idea in my head that nations can really act as ethically as they talk, if they are pressured by their peoples to do so.

Years ago I read an earlier mentioned newspaper column by the eminent Sydney Harris in which he described the world as a see saw with all peoples engaged in the activity of trying to be on top while the others are down. He noted that throughout history no matter who was on top they never acted better to those who were down than they were treated when they were there. Thus he concludes all societies in the aggregate are no better than any others.

I have cited that article many times in lectures in many parts of this globe. I suggest that Harris' contention is nearly irrefutable if one looks at history with any kind of objectivity. But I then add that because it has been true in the past is no sign that it has to be equally true in the future. I firmly believe that, and am enough of an optimist and ethnocentric to believe that it is my country, the USA, that has in its grasp the opportunity to change that pattern. Within our borders, in spite of a slip here and there including some pretty bad ones, I think we have come as close as is reasonable to expect, considering human frailties, to matching our rhetoric and our actions. We have indeed built a society within this nation for which we are rightly proud, one that tends to get better, fairer, and more exemplary with the passing ages.

But outside of our borders we have acted less in accord with our ideals. We are certainly not the worst of great nations in history in our foreign wheeling and dealing. In fact we have been considerably better than others who have held near equal influence and power. But we have still fallen far short of our own ideology. There have been and are too many ethical excesses done in the name of practicality, anger, puffery, security, or whatever.

But we are a young nation with a great potential for our young to rise to a higher standard than their predecessors. They are rich enough in mind, body and intellect to break that daisy chain of domination by the powerful over the powerless, if we will only let them do so. They are a powerful force. They turned this nation off its course in Vietnam, we can argue the right or wrong of that some other time. And they can, if we do not tarnish their ethics with too strong an implanted desire for wealth and comfort, reshape this nation's foreign policies so that while on top our nation will act as it should rather than as it could and too often has.

But they can do this wondrous, and indeed miraculous in history, selfless thing only if we give them the tools to make those proper judgements. And above all that means an inquiring mind, opportunity to learn about the other peoples on this planet, and an assurance that "different" does not always mean "less," "inferior," or "bad." And finally and most importantly they must be allowed to read, hear, question, and think about dissidence from the conventional wisdom.

Archibald McLeash wrote an article in the late '60s in which he predicted a return of the McCarthy period to American politics and letters by the mid-'80s. There are signs that he is right. Dissidence today in the USA is becoming unpopular, not because the USA is performing so ethically, but because those in power are discouraging dissent and encouraging a gutty patriotism that says, our nation may not be perfect but it sure better than all those others and if you do not agree—you are on the wrong side brother. If that system prevails, and I do not think it will, Sydney Harris will prove to be not only a top notch historian but a better prophet than I. Prove him wrong, will ya?

7/1/84

The SAVE US Solution

"The 'Save Us' solution is an idea we would like to see implemented." Four recent Latin American visitors to Montana and numerous Soviets and Americans have told me that lately. Each arguing that the solution was politically viable and vital to both world survival and development.

They were referring to a proposal of mine designed to begin to deal with what I consider the two most threatening problems on our globe: world poverty and the nuclear arms race. Problems which I suggest are inextricably intertwined.

Few would disagree that we have an arms race that seems out of control and a world poverty condition that is growing. But many, at least in the developed world, do not see the connection between the two. But informed people in the so-called Third World do. They see their debt grow less and less manageable as interest rates rise and recognize that the major reason for the interest rate increases is the US deficit, recently more than doubled by increased arms race efforts. Additionally, when they seek assistance from the rich they keep hearing the response that the wealthy would like to help but cannot because of the enormous cost of the arms race.

And at home in much of the Third World their own leaders, caught up in the world wide hysteria of insecurity, also deprive development projects, health and education, as they try to buy security, as they see the big boys doing it, by spending more and more on weaponry. The arms race and world hunger are part and parcel of the same problem, fear and misunderstandings among differing peoples, and they must be dealt with as one not two separate problems.

The US/Soviet relationship, the fear, distrust and apprehensions we foster toward one another are prime examples of this and may hold the key to resolving the problem worldwide.

In the 1940s the US and the USSR were also Christian and atheist, capitalist and communist; we distrusted and feared one another as much as now, although our rivalry was less threatening to all since then neither of us held the awesome power of hydrogen bombs. But we ignored our bitter differences and fears and worked together to defeat Hitler's threat to the world. We did so simply because we feared Nazi Germany even more than we feared one another, and the world as well as us benefited.

I suggest the time has come when we should both recognize that once again the world faces a danger greater than our mutual rivalry. It will not matter who is right or wrong if in the process of our near blind fear and hatred we destroy the world or become so calloused from our rivalry that we let the world around us sink into a mire. A mire much of it is now approaching with almost a billion going to bed hungry every night and millions of children dying with bloated stomachs, while their brothers and sisters suffer abnormal development due to lack of adequate nutrition.

Thus I propose the SOVUS SOLUTION. Pronounced "save us," the solution calls for a simultaneous joint Soviet/US action against both the arms race and world wide hunger and poverty. It begins with a mutual nuclear weapon freeze and a further agreement by both to set aside a portion of the funds each would save because of the freeze, say 50 percent, to begin a joint US/Soviet Commission. The Commission would be called either the SOVUS, OR SOVEUS COMMISSION. The "E" would be added if the European nuclear powers also joined the freeze and agreed to add funds and participate.

The Commission would be a non-governmental agency composed of equal numbers of experts from the participant nations and dedicated to eliminating world hunger by the year 2000. The SOVUS COMMISSION would allocate funds and send teams of experts (all composed of equal numbers of US and Soviets) to developing nations, regardless of their economic or political systems, to help plan and execute development projects. The project selection would be made independently by the Commissioners, who although appointed by their governments would hold non-revocable positions for set terms.

The SOVUS SOLUTION would thus blend arms control and development by freezing the nuclear arms race while finding ample monies for long term development projects. It is estimated the US alone would save about 100 billion dollars in the next five years if a freeze were implemented. Such sums could save millions more people if spent on water systems, schools, roads, new farming techniques, hospitals and sanitation rather than on more menacing warheads.

But equally importantly US and Soviet officials working together in significant numbers all over the globe on projects that could not but make all of the participants proud of their work and more tolerant and understanding of their co-workers, just might dissipate at least some of the differences that separate the US and the USSR today and help make real nuclear arms reductions possible five years or so hence. Media coverage of the activities of the joint teams could begin to create the kind of respect that might make arms reductions possible in the future that today are simply unacceptable.

The SOVUS SOLUTION offers no risks for it would leave the US and the USSR militarily no worse than at a standoff, as their and our chief military leaders recently have recognized publicly. SOVUS increases our foreign aid bill but at the expense of a bloated defense budget while still providing lower tax opportunities or a reduced deficit. It's an idea whose time is at hand. There are millions all over this globe who would feel far safer as a result of the SOVUS SOLUTION than by the addition of Midgetman or Peacemaker missile programs. All else has failed, let us try something that builds hope through cooperation rather than enmity through threats of holocaust, and might lead to rising living standards globally rather than a widening gap that daily increases the danger for all of us.

<table><tr><td>8/12/84</td><td></td></tr></table>

A Solution For Central America

My review of that cauldron called Central America (CA) has led me to conclude that US policies there will not bring about the desired result, and, in fact, are making the situation worse. Why? Because the USA is "mirror imaging" CA rather than recognizing the vast differences between their societies and ours.

No Central American nation, with the possible exception of Costa Rica, is similar to the US political and economic environment. Here, politicians chosen by the people run the government in accordance with the law. Power is diffused and the military is a professional soldiery subordinate to the political leadership. Private enterprise is regulated against instinctive excesses and labor has guaranteed rights and political clout. A large, relatively united middle class keeps pushing the nation toward the middle.

CA is different. It has evolved so that today its political leaders rule at the will of the military. Laws are irrelevant to those with power and the power is only slightly diffused, almost all of it held by the men who control the guns. CA's capitalism is nearly unfettered creating great wealth for a few but at a terrible cost to the many. The political middle has been polarized and battered and radicalization is the norm.

Based on that false, US-like image, our leaders seem to have concluded that all one needs to do to resolve the inequities in CA is to hold elections, vote out the inept or bad and replace them with new leaders who will pass new laws and make things better. But that seldom happens in CA, the people vote but most often out of fear and coercion. And on those relatively rare occasions when they elect someone other than the one preferred by those with the real power, the power brokers either steal the election, ignore the elected's efforts to take charge, or oust them via a military coup.

So what are we to do? If the current practices of combined economic and military aid plus pressures for decent reform cannot work, what other options do we have? A tempting alternative is to cut and run. If we are part of the problem why not pull out and let them settle it alone. But that temptation is misleading, and is not, as it may seem, a neutral position. The US has contributed mightily to the power of the CA oligarchy and the military. They kill with our guns and grow rich with our dollars. If we pull out we really side with the most radical elements in the ruling class and unleash them, without our constraints, to paint the isthmus red with the blood of the innocent.

I believe there is a third option. It would require a break both from the policies of the past and the oppositionists belief that almost any US involvement is bad. It is predicated on the conclusion that no solution is possible in CA as long as their military monopolizes the power, and that the US, since it helped create the miasma, has to actively seek to resolve it. I call it "Active Demilitarization" (AD).

AD would require the removal of the military leadership of failing nations like El Salvador and replacement of them, at first, with an outside security force, and later with a newly trained, officer corps subordinate to a new civilian leadership. All officers from Lt. Col. up would be retired and given a 60 day amnesty period. During the amnesty they and other potential criminals could leave the country scot free, but if they remain they would be subject to trial later for any crimes committed.

An OAS security force would occupy the country and guarantee security for a 3-6 month period. Both the rebel forces and the national military would be ordered to remain in their camps during the reconciliation period. Rebel leaders would be invited, with full security guarantees, to join in the constitutional process and the OAS supervised election. The resulting government, regardless of its eventual political alignment, would be pledged generous US and West European economic assistance as well as training for the new officer corps.

Under such an arrangement, the people of CA might vote quite differently than they have with their openly corrupt army providing security, and a government representing many factions, in cluding those the US would not like, is quite likely to emerge. The people of Algeria and Rhodesia voted one way when the gun holders were their white or foreign masters, but another quite different way when the situation was changed. The masses of CA deserve a similarly uncoerced electoral opportunity.

The US role in AD would be active, our support would be key to the agreement, probably requiring aid cutoff threats and tough negotiations with all major parties. We would pick up the major share of the initial costs and later economic assistance, but we should refuse to be involved in the security forces other than financially.

AD appears to me to be the only way to assist the people of CA to a better future. But sadly, it also appears no US leaders are ready to so drastically change our approach. If that remains so, CA's problems seem doomed to stay unresolved, and the likelihood of anti-US, radical leftist regimes to our south enhanced.

US Leaders Should Visit USSR

4/24/83

I am going to the USSR again to see what I can see and learn what I can about that nation and its people. Hindsight has shown us that ignorance and misunderstandings deserve a large share of the blame for past unsavory incidents between nations, and the best way to clear up those miscreances is via study and observation. Thus I believe the whole world benefits when Americans visit the USSR or vice versa.

I am not naive and do not expect to meet very many regular Ivans and to hear their innermost and honest thoughts. My trip, although designed to bring ordinary citizens into contact, will be orchestrated by the Communist party, but I lived in the USSR too long to believe that they can either blind my eyes or totally squelch the inherent Soviet citizen's desire to seek truth in spite of the curtain placed over it. My visit and all other such visits by "us" there or "them" here are, in my opinion, some of the most effective chinks we can drive into the chains that bind Soviet citizenry. And equally the most enlightening way for us to learn reality from the myths about the USSR that have plagued us too long and cost us too much in unreasonable fear and sweat generated dollars.

Of course the benefits increase with the influence of the individuals who make the visits. There is a bill before the US Congress with that in mind. It asks that a sum of money, far less than the costs of a jet fighter or a missile, be set aside to fund and encourage visits by US and Soviet officials to the other's nation. The fact is that about 80 percent of our current House members and some 58 percent of the Senators have never spent any time in that nation which so dramatically dominates much of our international actions and effects their decisions on how to spend US tax dollars. Most of them have visited many other lands and peoples but there seems to be an aversion to visiting the land of the Red almost as if it might contaminate them.

I would suggest that the bill should fund and urge that each member of the Congress, Cabinet, NSC staff, and our governors and key state legislators should spend a minimum of two weeks in the USSR and that at least part of their trip should be spent with a Soviet counterpart in his/her home region meeting with family, local constituency, and getting to know one another. Of course the counterpart would best be members of the Communist Party Politburo and Central Committee rather than members of the Supreme Soviet. In turn the Soviets would be asked to make a similar trip to the USA.

I firmly believe that such a program would serve world security far more than 100 times its cost in additional military expenditure by either side, and I speak from personal experience.I went to the USSR with profound misgivings and a certainty that their evil system directly threatened our very survival. I left, after a much longer stay than the one that would be possible for all of these key people, with a vastly changed outlook. And mine was not an atypical experience.

I did not become a fan of communism, but I did lose my fear of it. It became obvious that the concept is so overrated and the threat from it so overinflated that the fear tends to evaporate on the spot. Others have had similar experiences, almost always feeling less threatened when they view the inefficiency and backwardness, and also often surprised to find that the restrictions on freedom, although existent, are less blatant and restrictive than pictured in the West.

Of course not everyone reacts the same, but it is amazing how many do— including many very famous and influential visitors. Senators John Stennis and William Roth both went on record after their visits noting how they feared the USSR less than before they saw for themselves what Lenin's theories had finally wrought.

Unfortunately our president not only has not gone to the USSR but even refuses to meet Soviet leaders elsewhere, but that has not kept him from establishing a clear view about them apparently based on his perusal of the Reader's Digest.

I wish the president would join me in Alma Alta. We would both dislike a lot of what we would see there, but we would also both come home better informed and for us and you that would be a real plus. I suspect your Senators and Representatives would like to hear your thoughts on the bill providing for an exchange of visits between key Soviets and Americans. It has failed to pass on several occasions, perhaps due to the fear that voting for it makes one appear a Red sympathizer or a supporter of junkets. I think it is a marvelous idea that is well worth the cost. Let them know what you think.

1985

The Pros and Cons of Star Wars

Is it a cruel illusion or a beautiful dream, a defense or a debacle? That is the question many are asking about the concept called Star Wars or Strategic Defense Initiative (SDI). I thought it only fair to give you the best arguments of both sides and let you decide.

THE PROS

What can possibly be bad about trying to find a defense for nuclear weapons and perhaps rendering them obsolete as the president has asked this nation's scientific community to try and do? No one, no matter what you may hear to the contrary can tell you that it cannot be done. They can tell you the odds are against it, that it is highly unlikely, will be very expensive, will require an enormous number of breakthroughs that are not exactly on the horizon, and that there are more arguments against the likelihood than in favor of it, but they are lying to you if they say they know it cannot be done. For it just might.

After all we went to the moon, travel in space, cured many an awful disease, fly airplanes, drive cars, cross large bodies of waters on bridges and invented kleenex and Saran wrap. Most of those things were once considered impossible by the vast majority.

Now most scientists and military men currently engaged in scientific pursuit of SDI technology, do not really believe, in the near future at least, say the next 20 years, that we can render nuclear weapons obsolete and assure the safety of our populace and that of our allies in an all out attack by a determined nuclear foe. But in the process of researching the possibility they do believe that they might find that breakthrough that makes a perfect defense possible, and they are nearly sure they will discover how to put up a defense that will make it far less technologically possible than it is now for our opponents to think that they might be able to launch a surprise attack that could so cripple us, we would be unable to turn around and do unacceptable damage back to them. It is the fear of that possibility of a first strike, resulting from the so called "window of vulnerability" concerns, that encouraged people to look into the Star Wars scenarios in the first place.

Now our opponents, the "nay sayers," will try to tell you that in the process of developing this new defense we will scare the Soviets and upset the delicate balance of terror that now holds them in check. But one merely has to look at the USSR's historical record to conclude that it is better to have them scared of you than not. But they do not have to fear that if we develop the perfect, or even an improved defense, that we would launch a first strike attempt of our own against them. For we held a similar nuclear advantage over the USSR for several years in the '40s and early '50s when we could have pulled off a first strike and decapitated tens and possibly hundreds of major Soviet cities, but we proved we were not so inclined, even while they were enslaving their neighbors. They have nothing to fear from our Star Wars while we have something to gain. Either what the president suggested, or failing that, an improved and more secure insurance against a wild Soviet attempt for first strike.

Even that second alternative, which is most likely since there are so many things the Soviets could do to alter their attack force and enable it to sneak some through almost any defense, is well worth the effort. It will make me sleep better at night, decrease the likelihood of a catastrophic accidental nuclear war, offer better protection for my kids, and spur technology. How can one oppose that? Again, the critics of this endeavor to be free and independent, to protect our own rather than relying on Soviet rationality and goodness, suggest that arms control is a better answer than more weapons. But where and when has arms control worked? It is better to be self-reliant than deluded by concessions to people whose record of trust is questionable, if not obviously non-existent.

THE CONS

Technologically it is correct to say that we can not be absolutely sure that a SDI could not provide the perfect defense. But we do know that every so called perfect defense created in the past turned out to be a sham and a waste of money, and none of those were anywhere near as expensive or as complicated as the SDI task. So why not avoid all the fanfare that will probably only mislead the public with false hopes about being able to avoid their nuclear responsibilities, and continue on as we were before the president's controversial speech, doing SDI type research on a small and reasonable scale with prudence, minimal waste of coin and endeavor, yet steady progress toward more knowledge. As we are also sure the Soviets are doing.

But by elevating SDI to a national effort, a huge debate and multi-billion dollar effort, we have swung the whole affair into the realm of politics and perceptions, and in so doing created a nightmare that is guaranteed to spur the arms race. I say that with the assurance of an observer who watched the last strategic defense effort result in precisely that, an enormous increase in the number of offensive nuclear weapons. In fact increases far beyond their need, rhyme, or reason.

For you see we do have a precedence in this case. In the late '60s the USSR embarked on the development of a missile defensive system, then called ABM. I was in the USSR when that project was underway, and remember being impressed when Kosygin defended it by asking if it was not better to build weapons designed to defend than to attack. But American experts considered the Soviet action provocative and a deliberate means to upset the delicate balance of deterrence.

Thus, America reacted and its politicians, scientists and military all agreed that the best way to regain deterrence in the face of a Soviet effort to go defensive, not knowing how it might or might not work, was to assume the worse (that it would work) and figure out how to beat it. We did, and at a far cheaper cost than the price of building our own defenses, which we pursued for a short while after we were sure we had an offensive answer to the Soviet defense.

Later, we and the Soviets mutually concluded that our respective defensive endeavors were an expensive waste, simply not capable of protecting missiles or people and for all practical purposes we agreed to forget them, other than as protective research. But strangely neither of us then decided to faze out or even stop the continued momentum in our offensive forces that had been triggered by their defensive experiment. Thus, the offensive endeavors triggered have come to haunt us as we learn more about nuclear winters and other additional threats of giant scale nuclear wars.

So there may be technological reasons that will make the SDI a pipe dream, computer programs beyond our dreams and perhaps beyond our capability, the loss of human control over nuclear defenses and offenses, the ease of overcoming satellite launched lasers and particle beams, the threat of defensive weapons being used offensively, the current inability to generate the power needed to operate these exotic means of defense, etc. But they really are not the major issues here. The question is will pursuit of the SDI lead to a decrease in numbers of offensive nuclear weapons and therefore a safer world or not. And the answer seems obvious. A ringing, NO. We already have the ability to end humankind in an armageddon like spasm of nuclear terror and SDI will contribute more to making that capability grow than it would to diminish it. With deficits already crunching our ability to grow and meet the challenges of the '80s and '90s, why spend our coin and endeavors on something we have already been shown will produce the opposite of our goals. The answer is clear. Kill SDI and get on with limiting nuclear arms, not making more. We do not have to trust the Soviets to help us, we merely have to recognize that nuclear arms control is as much in their interest as ours.

There you are. Can you decide? You better for your life may depend on that decision. I think the arguments are obvious and I encourage my scientific friends to avoid SDI research as if it were a plague. Come to think of it, that's just what it is.

1985   Dare To Be Great

The book I hope to write some day asks the question, why does the USA act so differently beyond its borders than it does at home? For I have concluded that domestically we have a truly great nation. One that grows greater almost annually and by a definition of great that has nothing to do with national wealth and fancy inventions. At home we have created a society in which all have a chance to excel, succeed, be free and live at a reasonable, if not excessive, standard. Note I did not say that all of us have achieved those lofty possibilities, merely that we can, some with more difficulty than others, but clearly possible.

But outside our borders the American story is different and more spotty. Truly, even there we have had our moments of greatness, for example our treatment of Japan and Germany after WWII, the Marshall Plan that helped Europe back onto its feet. The billions of dollars and grain and medical supplies sent around the globe to peoples in need, the Peace Corps volunteers, who whatever their success, try to understand and help others at a sacrifice to their own living standard, etc. But in spite of these many gallant achievements, America's image in much of the world and to a growing number of the little people on this planet is lessening day by day. When once we were almost the dream and hope of almost all, a la the myth of the Statue of Liberty, something has happened and many look upon us today as no more worthy, and perhaps even less so, than the United Kingdom of the 19th century, the USSR of today, and the Rome of bygone days.

I do not like that and want to know why and how to change it. The "why" seems to me to be a result of our thinking that outside our borders we do not have to be as observant of laws, rights, and needs of others as we do at home. Partly, no doubt, because outside of our borders there is often not as much law, equity and concern for others, but partly too, I fear, because we fell prey to the wisdom of men like Morgenthau, T. Roosevelt, and Wilson who in their messianic way decided it was our duty to rule the world, convert it to our style, protect it from all threats, and mold it to our liking. All of this justified by the reality that out there, unlike in here, there are no rules except those made by the biggest and toughest, the meanest son of a bitch in the valley syndrome.

All of this has also come about partly because of our obsessive fear of communism and its number one advocate and our only real rival as a world force, the USSR. That fear has made us more aggressive, allowed us to act often without constraint simply because we feared if we did not, they would. Fear makes people do things they would not normally do and apparently effects nations similarly.

Thus, we turned to primary emphasis on the use of military force, or economic bribery, or even more often covert aggressive activity at the sub-war scale to shape events the way we preferred them. The revelations of the early '70s about US covert activities against Castro and other Central American states, in Africa and elsewhere shocked the nation for awhile, but after a few minor international setbacks, we were ready and eager to try them again.

I suggest that all of this comes about as a result of our political system, the inherent corruption that comes with power, and our lack of knowledge about the world we live in. And I would like to suggest a cure, some fundamental changes necessary that would encourage and enable us to act as responsibly beyond our borders as we do within them.

First, we should not allow our leaders, Congress, the president, his appointees, the military, the bureaucracy to feel that it is their duty to act, sometimes ignobly, to ensure our position in the world. We should pass a law that makes all domestic laws applicable to our international actions. US officials should be forbidden to commit an act that within our borders would be considered illegal. This to be overridden only by Congressional act after open debate. Such a law would make almost all of our covert activities illegal and lead eventually no doubt to the disbandment of covert organizations in the CIA, military and elsewhere.

Second, we should limit all US public officials to no more than one term at any level in our public service. This would almost eliminate professional politicians, bring more Americans into public service, and serve as a buffer against the inherent corruptive forces that come with too long an association with power. We have to avoid conditions that led one former president to conclude that if the president wants something done, it is legal.

Third, US military should be withdrawn from all nations where they are stationed in units. Deployments to other nations should have to be authorized by Congress, at least under the provisions of the warpowers act, after the fact in emergencies. It should be understood that US forces would only be used in other countries if those nations were actually under attack by foreign forces sponsored by another nation.

Fourth, the US should declare a nuclear freeze and promise to stop deployments or testing of any new nuclear warheads and call on the USSR and other nuclear powers to join us. We should proffer the "No First Use" policy with any nation that so agrees, and pursue arms control vigorously under the new atmosphere created by this non-production scenario. We should additionally pursue the Soveus Commission approach noted in the preceding pages.

Fifth, we should proffer no economic or military assistance to any nation, except in dire emergency, that does not show positive success in dealing with its poverty, hunger, and human rights deprivations.

It seems to me if the USA does not do these, or similar steps, we are doomed to dash our unique experience as a nation to a fate no different from that of all predecessor states this globe has experienced. That would be a terrible shame for we are something special, have done some glorious things and yet will doom them all to mediocrity if we do not dare to lead as different a path internationally as we have domestically. The hope is there, support is seen in the polls as our folk reject our policies supporting the Contras and opposing nuclear proliferation, but somehow it is always dashed as professional politicians seeking to perpetuate their careers mislead us, misunderstand us, and vote to survive rather than lead. Dare to be different America!!

1985

AC TO DC

In a recent article in *Atlantic* magazine the noted American historian, Frances Fitzgerald, described the major theme of America's post WW II foreign policy as "anti-communism" (AC). She concluded that this AC focused policy has been inadequate for the times and the realities and has led to American follies such as the war in Vietnam, our current policies toward tiny Nicaragua, and our support for almost any tin horn dictator who professes anti-communist rhetoric, a la Somoza, Mobutu, and Marcos, no matter how vile he might prove to be.

Fitzgerald explains this AC obsession by suggesting it lies deep in our religious fundamentalism and that those fundamentalist precepts play a far greater role in our basic beliefs and culture than they would reasonably be expected to do by the numbers of fundamentalists in our society. She focuses especially on those she calls the "premillennialists" who base their AC on the Biblical prophesies about the final battle to be waged on earth between the forces of Satan and GOD. She suggests they have asserted that the US (who else) is the nation of angels of those prophesies and that the USSR is the nation of Ros, the devil's armada. Thus, Fitzgerald suggests that for most American politicians and voters it is nearly impossible to view the USSR in any realistic terms since they have become the representatives of evil on this planet.

I do not know if the Fitzgerald explanation is the correct one, but I certainly concur that a so called anti-communism, but more accurately an anti-Sovietism, has dominated our nation's politics and foreign policy for decades and with often more deleterious than positive effects.

Therefore, I would like to offer an alternative focus for our foreign policies of the future. One that would give us more concrete measures of what policy to follow and one that is more flexible and would better explain and justify many of the policies we follow today.

In juxtaposition to the AC policy of the past I call this new one "Development Centered" (DC). Development Centered means simply that we judge our policies toward a particular nation state not on the trappings of its structure, but on the results. We concern ourselves not with whether a government calls itself or is labeled by others as capitalist, socialist, or communist; authoritarian, totalitarian, or democratic but instead by what happens to the vast majority of the peoples under its control. We look to see if the plight of the masses is improving, first economically and then politically and base our support, indifference, or opposition on the progress being made by the masses.

DC could correct the inconsistencies so obvious in our current policies and our AC rhetoric. For example today we sell the Soviets grain while calling them an evil empire, yet we have quite friendly relations with the PRC, another communist state, and have even offered to sell them weapons. But simultaneously we sponsor a war against Nicaragua, the least communist of the three, and have cut off all trade with them. Additionally our AC focus gives us no clear guideline on how to deal with South Africa, the Philippines, Zaire, and other staunchly anti-communist governments where living standards and political rights for their masses are deteriorating rapidly.

DC would solve those dilemmas. It would justify various forms of trade and relations with both the USSR and China, as it would with Yugoslavia, Hungary and other communist states by taking the focus off of their ism and transferring it to an evaluation of their living standard and a broad evaluation of the human rights of their populace, not as compared with the USA after its 200 hundred years of progress, but as compared with what those nations had before the current governments came on board. DC also suggests that human rights includes far more than just voting rights, independent judiciaries, etc. DC by its development focus suggests that important human rights lie in the area of available education, health facilities, housing, jobs, sanitation, etc.—things that are desperately lacking in far more nations of this globe than most Americans realize.

DC would demand toughened US policies toward the Philippines and Zaire, South Africa and other places where poverty is on the increase and systematic denials of human rights in the form of jobs, living conditions, etc., as well as in actual political power, are lacking and diminishing.

The AC policies of the past have tended to make America act, or at least appear to be like, most of its predecessor superpowers of history, nations bent more on increasing their power and in maintaining the status quo than in sharing their wealth and improving the lot of their less fortunate brothers and sisters. A switch from AC to DC gives the world a shot in the arm that suggests that the nation state patterns of the past can change and that people can care more about others than the dog eat dog world of our historical precedents would suggest is possible. I would like to see our nation be the one that breaks that mold forever, the one that admits that there are many ways to serve humankind, not just ours. The one that will be tolerant with success even where it differs from our approach, and the one that cares about humankind's progress rather than just our continued prosperity. It's time to switch from AC to DC!

BIBLIOGRAPHY

If these essays have whetted your appetite to learn more about the arms race, the USSR, US foreign policy, ethical issues of the day, etc., I offer the following as a list of my favorite books and periodicals which you might consider.

JOURNALS AND MAGAZINES

Foreign Affairs, Foreign Policy, World Policy Journal, Political Science Review, The Economist, The Nationa, National Review, Newsweek, Time, U.S. News and World Report, World Press Review, (this unique monthly provides only foreign viewpoints on major issues including cartoons, headlines and articles from all around the globe.), *The Wilson Quarterly, Alternatives, Orbis, Defense Analysis, Washington Post Weekly, New York Times Review of Books, Atlantic Monthly, Harpers, Business Week, Forbes, The New Yorker, Current History, Center for Defense Information, The Defense Monitor, Daedalus, Journal of Peace Research, Commentary, The Alternative, Department of State Bulletins, Scientific American, Science, Bulletin of Atomic Scientists, World Politics, Quest, Physics Today, the Air University Review, Reason, The Christian Century.*

BOOKS

Barnet, Richard, *The Economy of Death*, Atheneum, New York, 1969.

Barnet, Richard, *The Roots of War*, Penguin Books, New York,1972.

Bracken, Paul, *The Command and Control of Nuclear Forces*, CFR, 1985.

Castelli, Jim, *The Bishops and the Bomb*, Doubleday, Garden City, New York.

Cockburn, Andrew, *The Threat: Inside the Soviet Military Machine*, Random House, New York, 1983.

Cudsi, Alexander and Ali Dessouki, ed., *Islam and Power*, John Hopkins University Press, Baltimore, 1981.

Drell, Sidney, *Facing the Threat of Nuclear Weapons*, University of Washington Press, Seattle, 1983.

Ennes, James, Jr., *Assault On the Liberty*, Random House, NY, 1979.

Fallows, James, *National Defense*, Random House, New York, 1981.

Fitzgerald, Frances, *Fire in the Lake*, Random House, New York, 1972.

Green, Stephen, *Taking Sides*, Morrow and Co., New York, 1984.

Halberstam, David, *The Best and the Brightest*, Random House, NY, 1972.

Harvard Nuclear Study Group, *Living With Nuclear Weapons*, Harvard University Press, Cambridge, 1983.

Hoffmann, Stanley, *Primacy or World Order*, McGraw Hill Book Co., New York, 1978.

Hough, Jerry, and Merle Fainsod, *How the Soviet Union is Governed*, Harvard University Press, Cambridge MA, 1979.

Kennan, George, *The Nuclear Delusion*, Pantheon Books, New York, 1983.

Kwitny, Jonathan, *Endless Enemies: The Making of an Unfriendly World*, Congdon and Weed, Inc., New York, 1984.

Lappe, Frances Moore & Joseph Collins, *Food First, Beyond the Myth of Scarcity*, Ballantine Books, NY, 1977.

O'Keefe, Bernard J., *Nuclear Hostages*, Houghton Mifflin Co.,Boston, MA, 1983.

Pierre, Andrew, *The Global Politics of Arms Sales*, Princeton University Press, Princeton, NJ, 1982.

Scheer, Robert, *With Enough Shovels*, Random House, New York, 1983.

Schell, Jonathan, *The Fate of the Earth*, Avon, New York, 1982.

Smith, Hedrick, *The Russians*, The New York Times Book Co., New York, 1976.

Stoessinger, John G., *Why Nations Go to War*, St. Martin's Press, New York, 1982.

Shawcross, William, *Sideshow, Kissinger, Nixon and the Destruction of Cambodia*, Pocket Books, New York, 1979.

Talbot, Strobe, *Deadly Gambits*, Harper and Rowe, New York, 1984.

Talbot, Strobe, *Endgame*, Harper and Rowe, New York, 1980.

Timerman, Jacobo, *The Longest War: Israel in Lebanon*, Random House, 1982.

Tuchman, Barbara, *The Guns of August*, Bantam Books, New York, 1980.

Tuchman, Barbara, *The March of Folly*, Bantam Books, New York, 1984.

Turner, Stansfield, *Secrecy and Democracy, The CIA in Transition*, Houghton Mifflin Co., Boston, 1985.

Ulam, Adam B., *Dangerous Relations, The Soviet Union in World Politics, 1970-1982*, Oxford University Press, New York, 1983.

Weizman, Ezer, *The Battle for Peace*, Bantam Books, New York, 1981.

Woolsey, James, ed., *Nuclear Arms: Ethics, Strategy, Politics*, Institute for Contemporary Studies, San Francisco, 1984.

Yergin, Daniel, *Shattered Peace: The Origins of the Cold War and the National Security State*, Houghton Mifflin Co., Boston, 1977.

Zuckerman, Solly, *Nuclear Illusion and Reality*, Random House, New York, 1982.

ABOUT THE AUTHOR

Don Clark is a retired USAF colonel who currently is the Director of International Education at Montana State University where he also lectures on international relations for the Political Science Department.

During his military career Mr. Clark served in the Strategic Air Command, Headquarters Command, Air Training Command, Security Service and others. His assignments took him to the Soviet Union as an assistant military attache, to Japan, Turkey, East and West Europe, and visits to many of the nations and all of the areas of this world. He was the first USAF Research Fellow at the Fletcher School of Law and Diplomacy, Head of the Department of Military Environment (International Relations) at the Air Command and Staff College, and an Action Officer assigned to the Joint Staff in the Office of International Negotiations where he was involved either as an advisor to the Joint Chiefs, their representative to the National Security Council deliberations, or on US delegations dealing with such issues as SALT, MBFR, Law of the Sea, CSCE, Laws of Humanitarian Warfare, US/Soviet Rules of Naval Engagement, CCD, and others.

Mr. Clark has lectured at most of the military staff or war colleges, and at many universities, colleges, and before public and civic groups both in this nation and abroad on arms control issues, US Soviet relations, and other international issues. His articles have appeared in such diverse journals and magazines as THE AIR UNIVERSITY REVIEW, THE EDUCATIONAL RECORD, THE ALTERNATIVE, DEFENSE ANALYSIS, and SKIING. His weekly articles on international affairs appeared in four major Montana dailies and were selected for excerpt on Voice of America overseas news broadcasts.

Mr. Clark is married to the former Patricia Conway of Colorado. They have two sons and now reside in Bozeman, MT. Mr. Clark has been active in local public affairs serving as a board member and Annual Drive Chairman in the local United Way and Red Cross organizations as well as a member of the local Chamber of Commerce and its special organization the Green Coats.

His hobbies include skiing, tennis and other sports. He is a third degree (Sandan) in the sport of Judo which he studied at the Kodokan Judo College in Tokyo. He is very active in bringing foreigners to the USA and arranging for Americans to travel abroad.